THREONINE

FOOD SOURCES, FUNCTIONS AND HEALTH BENEFITS

BIOCHEMISTRY RESEARCH TRENDS

Additional books in this series can be found on Nova's website
under the Series tab.

Additional e-books in this series can be found on Nova's website
under the e-book tab.

BIOCHEMISTRY RESEARCH TRENDS

THREONINE

FOOD SOURCES, FUNCTIONS AND HEALTH BENEFITS

JACOB COLEMAN
EDITOR

New York

NOTICE TO THE READER

Library of Congress Cataloging-in-Publication Data

ISBN: 978-1-63482-554-2
Library of Congress Control Number: 2015936106

Published by Nova Science Publishers, Inc. † New York

CONTENTS

PREFACE

Threonine, one of proteinogenic as well as essential amino acids, is classified as polar and uncharged residue, which may play an important role in intermolecular interactions between protein molecules and small ligand molecules. This book includes chapters on the lysozyme of egg white, which contains seven threonine residues in the primary structure; the requirements of threonine for laying hens fed diets with increasing levels of threonine at variable and constant ratios; the role threonine may potentially play in intermolecular interactions towards ligands of small molecules and the highly mutated threonine-producer strains, and ways of optimization and intensification of its biosynthesis.

Chapter 1 - The highly productive strains, auxotrophic and regulatory mutants of microorganism producers and new unconventional substrates enables the improvement of essential amino acids biotechnology. The new productive producer strains receiving, study of their biological properties, mechanisms of adaptation to assimilate new substrates and choice of optimal growth medium has always remained relevant and requisite condition to improve the biotechnology of essential amino acids obtaining. This chapter summarizes recent research in these areas. Author's elaborations concerning the obtaining of highly mutated threonine-producer strain and ways of optimization and intensification of its biosynthesis are described in details.

In particular, results showing the findings of mutagen influence (UV irradiation) investigations on the cell culture *Brevibacterium flavum* TH 7 in the exponential stage of development are given in order to increase the biosynthetic activity of the producer, including an assessment of its lethal and mutagenic effects. It was shown that the resulting *Brevibacterium flavum* mutants, being resistant to β-oxynorvaline (2-amino-3-oxyvaleriane acid) as a selective agent, are carriers of two regulatory mutations that violate the retroinhibition of both homoserinedehydrogenase and aspartatekinase and excrete both threonine and lysine. It has been revealed that the threonine analogue β-oxynorvaline acts as retroinhibitor or corepressor of the natural metabolites synthesis, although it doesn't change them functionally. Consequently, only those cells can survive and form colonies at a minimal nutrient medium with antimetabolites which have broken mechanism of negative regulation of threonine biosynthesis and synthesize excess threonine as a result of this violation. The use of analog resistance as a genetic marker allows to select the high productive strains in the threonine accumulation. The obtaining of the mutant strain *Brevibacterium flavum* IMV B-7446 capable of accumulating 6 times more threonine compared with the parents (native) culture, by UV, was described. Analysis of the effect of different technological parameters for mutant strain cultivating on the synthesis of threonine and its stability was shown. The possibilities of intensify the threonine synthesis by putting growth factors in the culture medium of the mutant strain have been investigated. The nucleotide sequences of the 16S rRNA gene of both mutant and parents *Brevibacterium flavum* producer strains has been determined and dendrogram showing phylogenetic relationships of IMV B-7446 strain and related strains has been built.

Chapter 2 - This study is aimed to determine the requirement of threonine for laying hens fed diets with increasing levels of threonine at variable and constant ratios. Two sequential experiments were conducted: the first wherein diets contained increasing levels of threonine, with variable digestible threonine: digestible lysine ratios; based on the results, the second experiment evaluated the requirement of digestible threonine associated with a constant amino acids: lysine ratio. In the first experiment, diets were formulated based on corn and soybean meal, supplemented with industrial amino acids L-lysine, DL-methionine, L-tryptophan, L-isoleucine and L-valine to meet nutritional requirements for laying hens, except for threonine. Evaluated digestible-threonine levels were 0.446, 0.486, 0.526, 0.565, 0.605 and 0.645%. In the second experiment, treatments consisted of diets formulated with corn and soybean meal, supplemented with amino acids L-lysine, DL-methionine, L-tryptophan, L-arginine, L-isoleucine and L-valine to meet the nutritional requirements for laying hens. The ratios between the amino acids and lysine were 91, 23, 90, 83 and 75% for digestible methionine + cystine, tryptophan, valine, isoleucine, and threonine, respectively. Performance and egg quality data were evaluated. Based on these data, it was determined that the threonine requirement of laying hens is 0.597 and 0.610%, or 684 and 626 mg/layer/day, with variable ratio and constant ratio at 75%, respectively.

Chapter 3 - Threonine, one of proteinogenic as well as essential amino acids, is classified as polar and uncharged residue, which may play an important role in intermolecular interactions between protein molecules and small ligand molecules. Lysozyme of egg white contains seven threonine residues in primary structure, while HSA (human serum albumin), a certain transporting proteins in blood, of about 66 kDa, has some binding sites for external metal ions or metal complexes. The authors have prepared four new chiral salen-type Zn(II) complexes (**cyclo Zn**, **propane Zn**, **binaphtyl Zn**, and **Zn ntndd**) and characterized ^{1}H-NMR, fluorescence, UV-vis, and CD spectra. Among these Zn(II) complexes, **cyclo Zn** and **Zn tndd** are appropriate for the use of fluorescence proves against proteins because of intense emission when they are excited by UV light. Gradual spectral changes of UV-vis and CD spectra elucidated docking of these Zn(II) complexes toward lysozyme or HSA accompanying with deformation of secondary structures of proteins. Not only quenching fluorescence intensity by energy transfer but also Stren-Volmer analysis of fluorescence spectra suggested that the numbers of binding sites of **Zn tndd**-HSA complex are larger than **cyclo Zn**-HAS or **cyclo-Zn**-lysozyme complexes.

Chapter 4 - Serine, threonine, and alanine are majorly contained amino acid residues in laccase, a metalloprotein which reduces oxygen into water. In which threonine may potentially play a role in intermolecular interaction toward ligands of small molecules and catalytic group. Laccase are employed in typical biofuel cell cathodes with a mediator metal complex such as ferrocene.

In this paper, the authors report on oxygen reduction by laccase with other metal complexes known electron mediators in acetate buffer suspension and in carbon paste electrodes. Furthermore, in order to develop low-cost mediators, the authors prepared and tested some Cu(II) complexes, namely [Cu(phen)$_2$]$^{2+}$, [Cu(phen-derivative)$_2$]$^{2+}$, and [Cu(Schiff base)$_2$], which were characterized by means of elemental analysis, IR, UV-vis, and CD spectroscopy. Hybrid systems of the complexes and laccase were also prepared for comparison current and potentials of oxygen reduction.

In: Threonine
Editor: Jacob Coleman

ISBN: 978-1-63482-554-2
© 2015 Nova Science Publishers, Inc.

Chapter 1

THREONINE SYNTHESIS OF *BREVIBACTERIUM FLAVUM* MUTANT STRAIN

G. S. Andriiash, G. M. Zabolotna, A. F. Tkachenko, Ya. B. Blume and S. M. Shulga*
Institute for Food Biotechnology and Genomics,
National Academy of Sciences of Ukraine
Kyiv, Ukraine

ABSTRACT

The highly productive strains, auxotrophic and regulatory mutants of microorganism producers and new unconventional substrates enables the improvement of essential amino acids biotechnology. The new productive producer strains receiving, study of their biological properties, mechanisms of adaptation to assimilate new substrates and choice of optimal growth medium has always remained relevant and requisite condition to improve the biotechnology of essential amino acids obtaining. This chapter summarizes recent research in these areas. Author's elaborations concerning the obtaining of highly mutated threonine-producer strain and ways of optimization and intensification of its biosynthesis are described in details.

---

* E-mail: Shulga5@i.ua.

In particular, results showing the findings of mutagen influence (UV irradiation) investigations on the cell culture *Brevibacterium flavum* TH 7 in the exponential stage of development are given in order to increase the biosynthetic activity of the producer, including an assessment of its lethal and mutagenic effects. It was shown that the resulting *Brevibacterium flavum* mutants, being resistant to β-oxynorvaline (2-amino-3-oxyvaleriane acid) as a selective agent, are carriers of two regulatory mutations that violate the retroinhibition of both homoserinedehydrogenase and aspartatekinase and excrete both threonine and lysine. It has been revealed that the threonine analogue β-oxynorvaline acts as retroinhibitor or corepressor of the natural metabolites synthesis, although it doesn't change them functionally. Consequently, only those cells can survive and form colonies at a minimal nutrient medium with antimetabolites which have broken mechanism of negative regulation of threonine biosynthesis and synthesize excess threonine as a result of this violation. The use of analog resistance as a genetic marker allows to select the high productive strains in the threonine accumulation. The obtaining of the mutant strain *Brevibacterium flavum* IMV B-7446 capable of accumulating 6 times more threonine compared with the parents (native) culture, by UV, was described. Analysis of the effect of different technological parameters for mutant strain cultivating on the synthesis of threonine and its stability was shown. The possibilities of intensify the threonine synthesis by putting growth factors in the culture medium of the mutant strain have been investigated. The nucleotide sequences of the 16S rRNA gene of both mutant and parents *Brevibacterium flavum* producer strains has been determined and dendrogram showing phylogenetic relationships of IMV B-7446 strain and related strains has been built.

Keywords: threonine, UV mutagenesis, β-oxynorvaline, mutant strain producer

INTRODUCTION

Threonine (α-amino-β-hydroxybutyric acid) is a hydroxyamino acid, one of the twenty standard amino acids. Threonine is essential amino acid and the second limiting amino acid after lysine. It cannot be synthesized in the body and has to enter the body only with food. It participates in metabolism of fat, collagen, elastin, and supports protein metabolism in the body. Threonine stimulates immunity, conduces to formation of antibodies and controls digestion of feed. Threonine has effect on growth of skeleton muscles, synthesis of immune proteins, digestive enzymes, glycerol. Threonine is a lipotropic substance which prevents fat accumulation in the liver and participates in synthesis of purines [1, 2]. Threonine is used as feed supplement to ration of a number of animals.

L-threonine is mainly obtained by microbiological method therefore process intensification, as rule, has to be enabled at the expense of the rise in producer strain productivity [3-6]. Bacterial cultures, higher plants, some of the algae feature threonine biosynthesis process through α-diaminopimelic acid (DAP-process) which starts from asparaginic acid (figure 1).

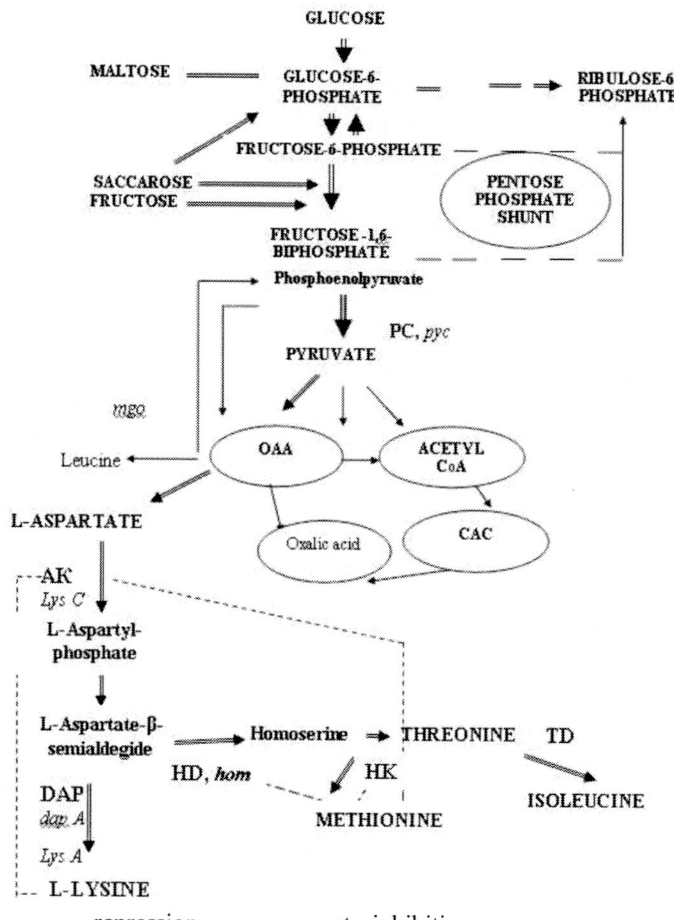

Note: - - - - - - - - - repression, - - - - - - - - - - - retroinhibition.

Figure 1. Threonine synthesis through the DAP-pathway.

In addition to lysine and threonine, the branched biosynthetic scheme is also present in formation of methionine and isoleucine [7]. Microorganisms synthesize each of the amino acids in certain quantity and control over biosynthesis speed of each amino acid is reproduced on feedback basis both on the level of genes responsible for synthesis of relevant enzymes (repression), and on the level of enzymes themselves which, when affected by excess of the amino acids formed, are capable of changing their activity (retroinhibition). Such control excludes overproduction of amino acids, and their release from the cell is possible only with the microorganisms whose regulation system is impaired.

Biosynthesis of aspartate family amino acids is controlled on the level of the first enzyme of β-aspartokinase (AK). Lysine and threonine producers (glutamate-producing corynebacteria *Corynebacterium glutamicum, Brevibacterium flavum*) have only one AK, whose activity is regulated by consistent inhibition on feedback basis by threonine and lysine [8]. To attain to overproduction of certain amino acids, such control mechanism in their synthesis has to be either bypassed or changed. Changes in amino acid synthesis control mechanism are made using present-day methods of genetic engineering.

The result is that auxotrophic and regulatory mutants are obtained [8, 9]. The mutant which is resistant to threonine analog (α-amino-β-oxyvaleric acid or norvalin) synthesizes threonine in excessive quantity. Homoserine dehydrogenase (HD) and homoserine kinase (HK) which take part in threonine synthesis are «excluded» by L-methionine, therefore methionine auxotrophs give higher concentration of L-threonine [3, 7, 8, 10, 11].

This chapter considers the process for obtaining of threonine producer strain with increased synthetic activity. Respectively, here we described the investigation of strain regulatory auxotrophy, selection of the producer for enhancement, performing of UV mutagenesis, testing of producer biosynthetic capability, phylogenetic analysis of the obtained producer and determining of optimal cultivating parameters.

1. PRODUCERS AND PATHWAYS OF THREONINE SYNTHESIS INTENSIFICATION

As a rule, commercial producers obtain L-threonine using traditional methods of stepwise selection. The obtained commercial mutant strains

Saccharomyces cerevisiae synthesized in sugar-containing mediums 40 times as much threonine as the stock strain [12, 13]; *Serratia marcescens* strains produced threonine with concentration of 13 g/dm^3 in the medium with 10% glucose [14]; *Methylobacillus glicogenes* strains produced threonine with concentration of 13,5 g/dm^3 in synthetic methanol [15].

Corynebacterium glutamicum, Brevibacterium flavum, Brevibacterium lactofermentum, nonpathogenic gram-positive corynebacteria producers of threonine excreted from natural sources and belonging to several genera (*Brevibacterium, Corynebacterium, Arthrobacter* , etc.) and species, form at the moment the group with various generic and specific names. The studies showed identity of biochemical features and high level of DNA homology in many representatives of these genera. Due to genetic proximity of these microorganisms and complex of taxonomic features of glutamate-producing corynebacteria strain, they may be ascribed to one genus of *Corynebacterium* [3, 8].

Achievements of molecular genetics and genetic engineering created real prerequisites for strain designing which include cloning of individual genes, their amplification in multi-copy plasmids, substitution of promoters to raise gene expression, site-specific mutagenesis. For gram-positive non-sporogenous corynebacteria and brevibacteria such manipulations are difficult because the majority of plasmid vectors are replicated only in gram-negative microorganisms. Applying of genetic engineering methods to brevibacteria becomes complicated also due to methods of foreign DNA introduction into cells [3].

To enhance threonine producer, *E.coli* optional anaerobe was used, and with the help of genetic recombination recombinant producer of threonine was obtained. *E. coli* MG 442 strain was transformed by plasmid with the gene of pyruvate carboxylase. Introduction of gene of pyruvate carboxylase (*pyc* gene) into *E. coli,* increased amino acid production [3, 5, 16, 17].

Significant rise in biosynthesis was achieved with the help of manipulation with *thrE, rhtA, rht B, rhtC* genes responsible for threonine transport [8,18]. The producer had inactivated threonine dehydratase and produced up to 100 g/dm^3 threonine in 36 hours of fermentation.

Search for natural corynebacteria plasmids gave the result that from *Brevibacterium lactofermentum* wild type strain identical cryptic plasmids were excreted with molecular mass 4,4 kbp, with unique restriction sites for several restrictases – *pAM330, pBL1, pWS101, pX18, pGX190* [17, 18].

Introduction of selective markers (genes of resistance to antibiotics) into cryptic plasmids creates a large number of double-replicated shuttle vectors.

RNA-polymerase of corynebacteria is capable of identifying various promoters of gram-positive and gram-negative bacteria. But resistance to antibiotics is much lower in corynebacteria cells compared to the cells of *E. coli* and bacilli.

Corynebacteria strain capability of recombination plays also positive role. In case of spontaneous formation, large deletion vectors having lost a part of mass preserve selective markers and become stable. When setting up systems for corynebacteria gene cloning, phages are also used which bring about lysis in commercial conditions. A cosmid vector was created on the basis of *F/A1* phage, which has cohesive ends to *pAJ43* plasmid vector [3]. This vector is «packed into» phage's head *in vivo* and transmitted during phage infection.

Cloning of *Corynebacterium glutamicum* wild type strain GD gene in *pCE152* vector made it possible to obtain *pChom9* hybrid plasmid. In the cells which carry hybrid plasmid HD and HK activity increased by 17 times. To prevent threonine HD inhibition, plasmid mutagenesis was conducted *in vivo*. From the resistant clones *pChom93* plasmid was excreted which encoded the mutant HD. Sensitivity to threonine in this enzyme was decreased by 400 times. The effect was produced when the plasmid with mutant gene was introduced into the strain with defective threonine desaminase – enzyme which catalyzes transformation of threonine into isoleucine. Recombinant strain with *pChom93* plasmid produced 17 g/dm^3 threonine in test tubes and 52 g/dm^3 in laboratory fermenter (29% sugar conversion) [3, 8, 18].

Synthesis intensification may be achieved by direct influence on producer strains to obtain strains with higher productivity and also by extension of substrate application spectrum, optimization of cultivating conditions, enhancement of mass exchange processes and process equipment [3, 4, 8 19]. Methods and pathways of biosynthesis intensification are presented in figure 2. For a substrate, producer microorganisms may use certain chemical compounds, natural polymers and food industry and agricultural wastes – molasses, sucrose, urea or ammonia, ammonium diphosphate, phosphates, corn extract, etc. [4, 19].

Use of auxotrophic microorganisms in threonine biotechnology is also connected with biotechnological techniques of synthesis increase (figure 2).

One of the efficient methods to obtain high-productivity strains of microorganisms is selection of mutant clones under action of various mutagens [20, 21]. Spontaneous mutants are usually detected with frequency 10^{-6}- 10^{-8}. This frequency can be significantly increased by treatment of cells with mutagens. The microorganisms with impaired regulation system are obtained by excreting them from natural substrates and by action of various mutagens (chemical, physical, biological) on microorganism culture with subsequent selection of the strain with preset features.

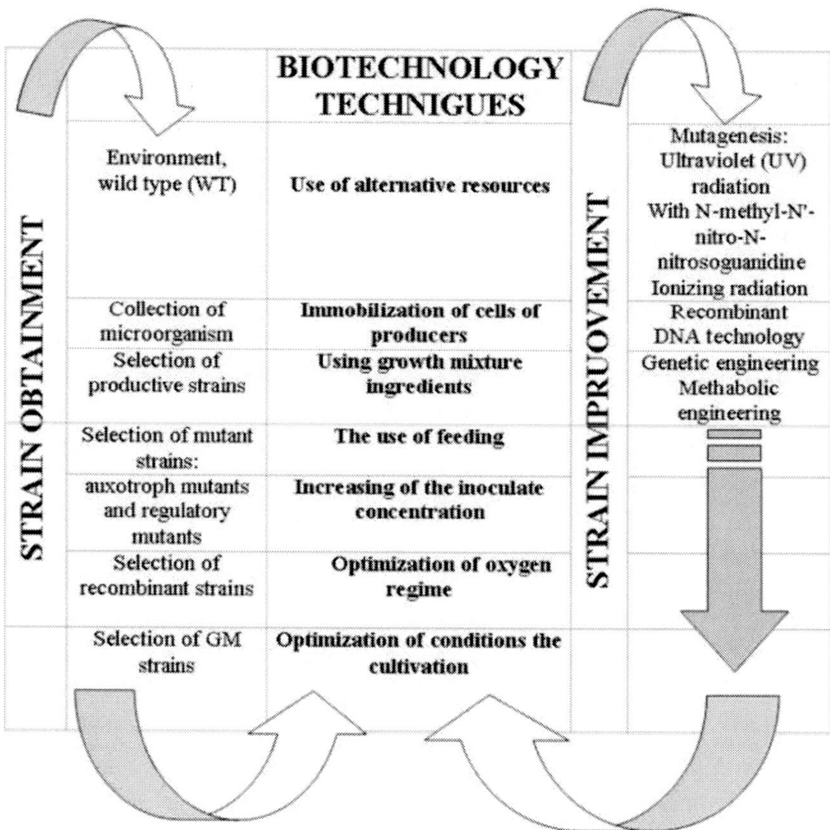

Figure 2. Pathways of threonine biosynthesis intensify.

Examination of enzymatic processes in threonine producers made it possible to optimize the composition of medium, define optimal concentrations of carbon and nitrogen sources for nutrient medium, determine impact of medium components on biotechnological performance and select appropriate cultivating conditions to obtain maximum concentration of biomass and threonine [8].

To obtain threonine producer mutants from corynebacteria family, selection of regulatory mutants is used whose HD is not sensitive to threonine. As a selective agent, threonine analog - β-oxynorvalin (2-amino-3-oxyvaleric acid) is used. *Brevibacterium flavum* mutants are resistant to β-oxynorvalin and have two regulatory mutations which impair retroinhibition of both HD and AK. With such strains threonine and lysine were excreted into medium simultaneously. It is very seldom that one can manage to excrete the mutants in which only HD is desensitized and which accumulate in the medium 10-12 g/dm^3 threonine without lysine impurities [3, 7, 8, 10].

Corynebacteria genes are cloned in two pathways. The first pathway is heterologous complementation of genetically and biochemically well-known auxotrophs of corynebacteria chromosome DNA *E. coli* using plasmid or cosmid vector.

For *E. coli* culture it is established that the genes which encode synthesis of AK, HK and threonine synthase (TS) constitute threonine operon. As a result of such arrangement of structural genes, *E. coli* culture may be used to obtain threonine overproducers. Based on the mutants excreted from *E. coli* K-12 stock culture, the threonine producer was obtained which does not require introduction of any amino acids into the medium.

The second pathway is direct cloning of corynebacteria genes by complementation of auxotrophs of these microorganisms [3, 8, 22-24]. In the study [17], the genes were cloned which encode threonine biosynthesis enzymes (homoserine dehydrogenase, homoserine kinase) in *C. glutamicum* and *B. lactofermentum*. *E. coli* threonine operon genes were used for enhancement of *B. flavum* threonine producer. *pCEM300* shuttle vector and new *pEC71* shuttle vector were used and plasmids were obtained which carried *E. coli* threonine synthesis genes.

After chemical mutagenesis depending on resistance to norvalin the mutants with recombinant plasmids were excreted. *B flavum* mutant strain had cloned genes *thrA, thrB*. The strain produced up to 12 g/dm^3 threonine in 48 hours of fermentation.

Cloned in the study [25] were *hom-thrB* threonine operon genes and *thrC C. glutamicum* ATCC 13032 gene and *hom* gene with resistance to threonine inhibition, which were cloned respectively with *hom FBR-thrB* operon *C. glutamicum* DM 368-3 and in combination with *E. coli* / *C. glutamicum* genes. Enzymatic activity of homoserine dehydrogenase, homoserine kinase and threonine synthase in recombinant strains as a rule increases by 8-20 times. It should be noted that for *C. glutamicum* wild type strains increase in threonine gene dose did not lead to increase of threonine production. Overexpression of *thrC* gene or in combination with *hom FBR* and *thrB* genes did not result in overproduction of threonine or lysine in recombinant strains either.

2. OBTAINING OF THREONINE MUTANT PRODUCER STRAINS

2.1. Objects, Materials and Methods of Studies

Objects of the study were producer strains of essential amino acids *Brevibacterium flavum* TH-7 and *Brevibacterium flavum* IMB B-7446 from «Collection of microorganism strains and lines of plant lines for food and agricultural biotechnology» of Institute for Food Biotechnology and Genomics, National Academy of Science of Ukraine.

Conditions of cultivating and mediums. Used in growing of producer strains were complete nutrient mediums – meat-peptone agar (MPA) and rich meat-peptone agar (MPA_{rich}) [4, 9]. Culture purity and productivity was checked once a year (museum cultures).

To determine strain auxotrophy and to conduct mutagenesis, bacterial suspension was used prepared as follows: two-day culture was sampled from MPA_{rich} dashed cultures and diluted in sterile saline solution to concentration of 1×10^5 colony-forming units (CFUs)/dm^3, which corresponded to 0,5 optical density (OD). OD was measured in cuvettes with d=5,0 mm at 440 nm wavelength with the help of photoelectrocolorimeter (of KFK-3 model).

The obtained inoculate was aseptically transferred to: a) complete medium (MPArich), b) minimal medium (MM) (glucose or sucrose – 3,0%, $(NH_4)_2SO_4$ – 1,0 %, K_2HPO_4 – 0,2%, $MgSO_4$ $x7H_2O$ – 0,04%), c) MM with the studied amino acid (MM + leucine or homoserine) and studied antimetabolite (MM + β-oxynorvalin, d) molasses medium. Composition of molasses medium: (g/dm^3): molasses – 160,0, corn extract – 40,0; $(NH_4)_2SO_4$ – 15,0; KH_2PO_4 – 0,5; K_2HPO_4 – 0,5; $MgSO_4$ $x7H_2O$ – 0,25; biotin – $3,0 \times 10^{-4}$; leucine – $2,0 \times 10^{-4}$; $FeSO_4$ $x7H_2O$ – 0,01; $MnSO_4$ xH_2O – 0,01; $ZnSO_4$ $x7H_2O$ – 0,001; $CuSO_4$ – 0,2; $NiCl_2$ – 0,02.

After sterilization sterile chalk was introduced in quantity of 10 g/dm^3 to create medium buffering in the course of bacteria metabolism. Submerged cultivation took place in 0,25 dm^3 Erlenmeyer flasks with 0,03 dm^3 nutrient medium at temperature 31±1°C and at 240 $min.^{-1}$ in "BIOSAN" ES-20 shaker incubator for three-four days. The cultivating process was monitored directly using microscopy of live drugs and biochemical analyses of culture liquid (CL).

The growth of threonine producer strains was judged from the fact of growth and formation of pigment (solid nutrient mediums); in liquid nutrient mediums – from measurement of cell concentration in culture liquid by optical density (OD); from medium pH value change variations, from utilization of sugars – by resorcinol method [26]. Quantity of synthesized target amino acids was measured using AAA-400 amino acid analyzer (Ingos).

Auxotrophy and sensitivity to antibiotics were studied in accordance with the methods [4, 8] modified for bacterial producers. As complete nutrient medium and positive control, MPA_{rich} was used, as negative control – MM. Glucose, sucrose and amino acids were sterilized separately and introduced into MM. Solutions of amino acids (amino acid test portion with weight of $0,19g$ in $0,025dm^3$ distilled water) were sterilized for 15 min. at 49 kPa pressure. Sterile solutions of amino acids with volume of $0,004$ dm^3 were introduced into molten MM ($0,05dm^3$), mixed and distributed in Petri dishes. Incubation was run at temperature $31\pm1°C$ for 2-3 days.

Content of various medium components was varied depending on the task assigned. Carbon sources – glucose, fructose, sucrose – were introduced into medium on the basis of 40 g/dm^3. Concentration of Na_2CO_3 varied from 0,1 to 0,5 g/dm^3; proline from 0,4 to 2,0 mg/dm^3; thiamin HCl $1\text{-}5x10^{-3}$ g/dm^3; biotin $1\text{-}5x10^{-4}$ g/dm^3; yeast extract 0,5-2,5 g/dm^3; isoleucine and methionine from 0,1 to 0,5 $\mu g/dm^3$.

To estimate efficiency of alternative carbon sources, beet molasses, milk serum and synthetic medium were used. The nutrient molasses medium of the following composition was investigated, % per 0,1 dm^3 of tap water: molasses – 16,0; $(NH_4)_2SO_4$ – 1,5; KH_2PO_4 – 0,05; K_2HPO_4 – 0,05; yeast extract – 0,25; and serum medium of the following composition, % per 100 ml of milk serum: glucose – 8,0; $(NH_4)_2SO_4$ – 1,5; peptone – 0,1; yeast extract – 0,25. Added to the mediums were also amino acids in quantity of 2,5 $\mu g/0,1$ dm^3: methionine, lysine, isoleucine, or biotin in quantity of 200 μg per 0,1 dm^3.

Reagents. β-oxynorvalin (Sigma) threonine analog, kit of indispensable amino acids (Shanghai Seebio Biotech. Inc.), antibiotic disc set («Aspekt») were used in the study. The reagents for biochemical and electrophoretic investigations were prepared on the basis of purified deionized water («DIRECT Q3» system, «Millipore»). DNAs were excreted by relevant «Fermentas» reagent kit. For electrophoresis agarose, ethidium bromide (basic solution with concentration 10 g/dm^3) and bromophenol blue (all from Sigma) were used.

In accordance with the methodology set forth in the study [20] mutagenesis was made by UV irradiation (two «Phillips» lamps 30 W each were used, $\lambda=254$ nm, distance to the object of irradiation – 0,12 m) of bacterial suspensions at room temperature for 60 – 720 sec. every 60 sec.. The irradiated bacterial suspension was inoculated at various dilutions (from the suspension initial concentration to 10^{-6} dilution) into minimal medium with amino acids and target amino acid analog [8].

Incubation took place in thermostat at temperature $31\pm1°C$ for three days. All the colonies which grew in MM with amino acids and in MM with amino acid analogs and which were auxotrophic to leucine and homoserine, were checked for production of amino acids. The most productive clones were picked out for further steps of irradiation and investigations.

DNA isolation. To isolate DNA bacteria cells were taken from one-day culture obtained in MPB_{rich} at temperature $31\pm1^{0}C$ under conditions of aeration at 220 min.$^{-1}$. DNA isolation ran following the standard procedure for gram-positive bacteria according to [27]. For more efficient cell lysis, 1% lysozyme (10 mg/ml) was added. The isolated DNA was studied using horizontal electrophoresis and PCR [28-30]. Electrophoretic separation of the isolated DNA took place in 1% agarose gel in tris acetate buffer system. Molecular mass of DNA fragments was measured from their electrophoretic mobility using as markers 1kb – DNA marker (1kb Fermentas SM1163) [28].

PCR conditions. Amplification of 16S rRNA gene was performed with the help of all-purpose bacterial primers 27f and 1492r (27F 5'-AGA GTT TGA TGG CTC AG-3'; 1492r 5'-TAC GGT TAC CTT GTT ACG ACT T-3'). For PCR «Mastercycler personal 5332» (Eppendorf) cycler was used. The reaction mixture consisted of single PCR-buffer with ammonium sulfate, 0,2 µM of relevant primers, 200 µM of each of deoxynucleotidtriphosphates, 0,5 units of Taq-polymerase (Fermentas), 2,0 mM of magnesium chloride, 10-50 ng of DNA-assay. Total volume of the reaction mixture: $20x10^{-3}$ dm^{3}.

Amplification conditions: initial denaturation at $T=95^{0}C$ – 3 min.; 32 amplification cycles ($T=94$ ^{0}C – 30 s, $T=57$ ^{0}C – 45 s, $T=72$ ^{0}C – 30 s); final elongation took place at $T=72$ ^{0}C for 5 min. [31]. Electrophoretic separation of the obtained amplification products was made in tris-acetate buffer. The obtained fragment was extracted from agarose gel using «Macherey-Nagel NucleoSpin Extract» kit in accordance with manufacturer's instruction and sequenced in automatic sequencer.

After gene amplification the nucleotide sequence of the obtained amplicon was determined using «ABI PRISM 310 Genetic Analyser» sequencer (Applied Biosystems). The resulting sequencing contig was obtained by comparison of direct and reverse complement sequences using CLC Main Workbench software (CLC bio). Homological sequences were picked up from «GenBank» database [32].

Comparative analysis of nucleotide sequences. Phylogenetic analysis. From «GenBank» database 16S rRNA gene sequences were picked up belonging to various representatives of brevibacteria genus which have the highest level of nucleotide similarity to sequenced fragments of 16S rRNA gene of investigated lysine producer strains. To clarify the systematic position of the investigated strains in relation to related ones, relevant nucleotide sequences were aligned in ClustalW program [33] and tree diagram of phylogenetic relationships was drawn up. Phylogenetic analysis was made in MEGA6 program [34, 35].

Statistic processing of data was performed with the help of Microsoft Excel. All the experiments were carried out in 3 replications. Difference between two average values was considered probable at $p<0,05$.

2.2. UV-Mutagenesis

In the population of microorganisms after UV irradiation, equally probable was occurrence of both mutants and revertants. It was therefore important after irradiation of producer strains to detect the strains with increased synthesis of threonine.

One of the methods to obtain threonine producer is selection of regulatory mutants whose HD is insensitive to threonine. As selective agents threonine analog β-oxynorvalin was used. The mutants resistant to norvalin had two regulatory mutations which impaired retroinhibition of both HD and AK. Use of analog resistance as genetic marker made it possible to choose the most productive strains in relation to the target amino acid [7].

Threonine analog β-oxynorvalin acted as retroinhibitor or co-suppressor of natural metabolite synthesis but it failed to substitute them functionally. Therefore in minimal medium with antimetabolite the colonies of those only cells survived and were formed which had impaired mechanism of amino acid biosynthesis negative regulation and which, because of this impairment, synthesized threonine to excess [21].

According to the mutagenesis scheme (figure 3) at the first stage of investigation the effect of mutagenic factors on cell viability was determined. For reference, relevant dilutions of bacteria unirradiated suspensions were taken. Cell viability was assessed before and after UV irradiation of bacterial suspension. Number of surviving cells in the population was measured from the quantity of the colonies formed in MPA_{rich} and depending on UV exposure duration. For reference the unirradiated bacterial suspension was used.

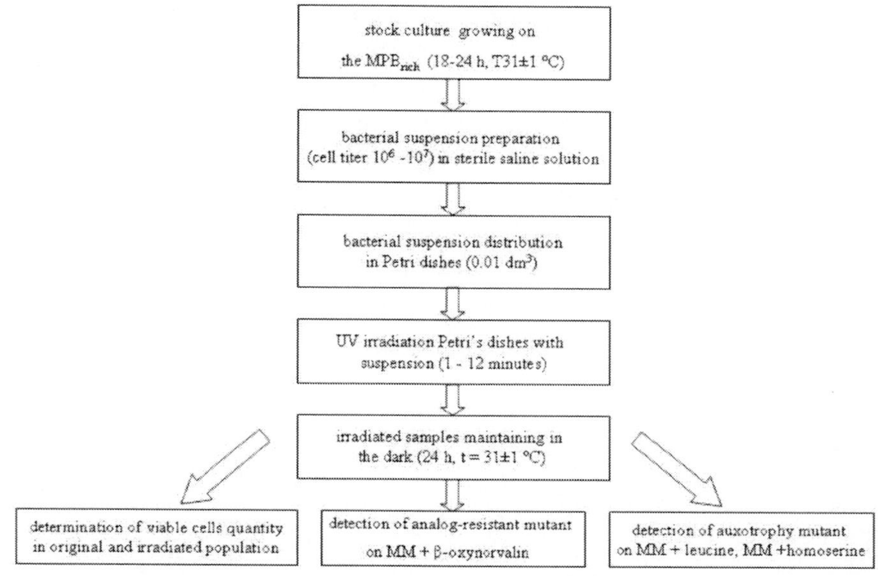

Figure 3. General scheme of investigations on mutagenesis of threonine producers.

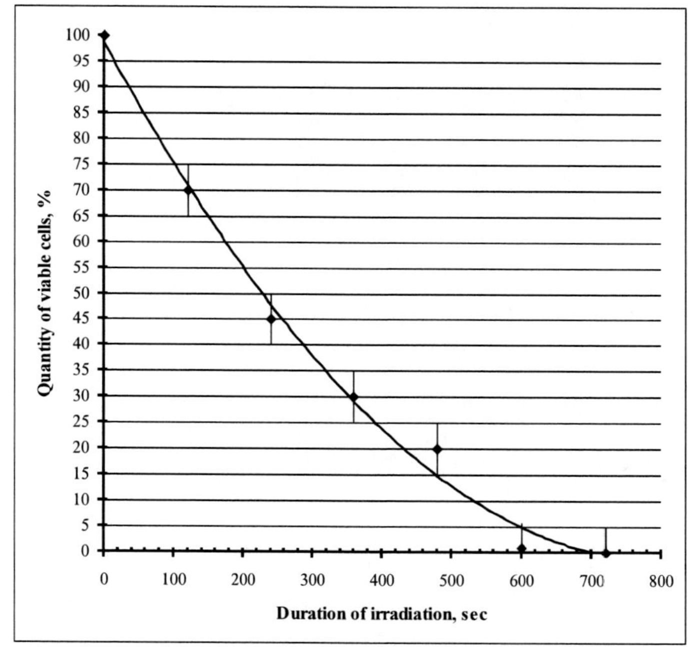

Note: 100% CFU in the non-irradiated bacterial suspension served as a control.

Figure 4. Cell viability of *B. flavum* TH7 under UV irradiation.

Probability of mutations in bacterial populations is as much higher as higher the number of cells which die is, there had to be however a certain number of live cells left for selection of mutants (about 1%) [21, 28].

Lethal dose (LD) and exposure time to get 1% of live cells differed from strain to strain (figure 4).

Lethal dose for *B. flavum* TH7 was irradiation for 12 min., and optimal time for mutagenesis was 10 min.

Among all the obtained clones were picked up and checked for biosynthesis of the target amino acids (figure 5).

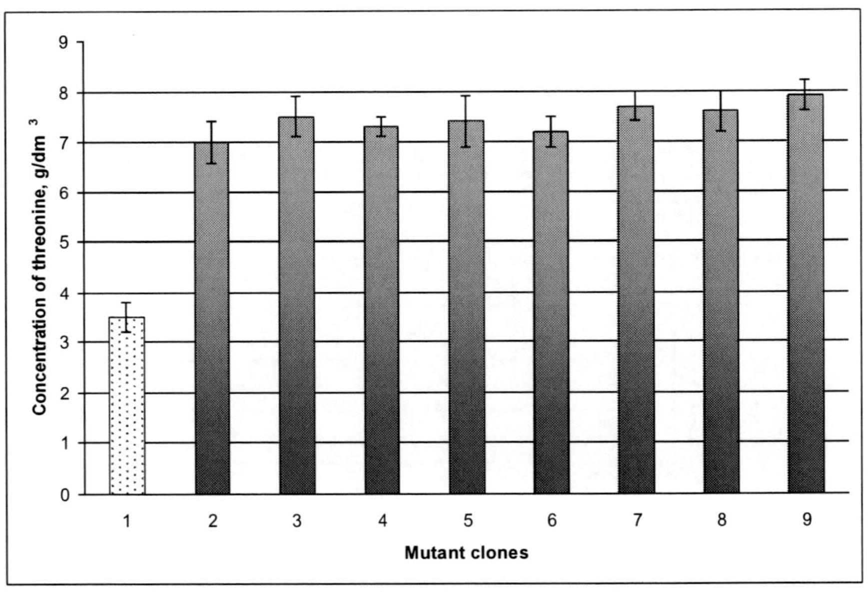

Note: 1 – strain before UV radiation; 2-9 – derived UV mutants; synthesis rate of threonine before radiation served as a control.

Figure 5. Threonine accumulation of mutant clones *B. flavum*.

The mutant strains showed increase in threonine synthesis level more than 2 times.

2.3. Analysis of Auxotrophy and Strain Identification

Resistance of amino acid analogs may be brought about by the mutations which simply block their entry into the cell. Several steps of selection were performed using gradual rise in analog concentration (from 0,25 mg/dm^3 to 0,4 mg/dm^3 β-oxynorvalin). Irradiated bacterial suspension was inoculated by submerged method in MPA$_{rich}$ (to find out cell titer), MM with leucine or homoserine and with β-oxynorvalin (to select mutants by criteria of auxotrophy and analog-resistance). The results are given in table 1. The mutants were selected by criteria of auxotrophy and resistance to β-oxynorvalin for threonine producers. β-oxynorvalin-resistant threonine producers were obtained with mutation frequency of $(1,4\pm0,2)\text{x}10^{-3}$.

Table 1. Forming of *Brevibacterium flavum* mutant strains by UV

Dilution	CFU mutant clones on MM+		
	leucine	homoserine	β-oxynorvalin
10	0	0	0
10^2	0	0	0
10^3	$(2,5\pm0,3)\text{x}10^{-3}$	$(2,1\pm0,3)\text{x}10^{-3}$	$(1,4\pm0,2)\text{x}10^{-3}$
10^4	$(3,3\pm0,2)\text{x}10^{-4}$	$(3,4\pm0,3)\text{x}10^{-3}$	$(1,5\pm0,3)\text{x}10^{-4}$
10^5	$(1,8\pm0,3)\text{x}10^{-5}$	$(5,2\pm0,2)\text{x}10^{-4}$	$(1,8\pm0,3)\text{x}10^{-5}$

The stock culture and obtained mutant strain were checked for sensitivity to antibiotics to establish the genetic markers. The results are given in table 2.

Table 2 Sensitivity of stock and mutant strains to antibiotics

Antibiotics	Straine-producers		
	Brevibacter ium flavum	threonine mutant production strain	resistant of mutant strain to β-oxynorvalin
Azithromycin	S	S	S
Ampicillin	S	S	S
Ceftriaxone	S	S	S
Benzylpenicillin	S	S	S
Gentamicin	S	S	S
Tetracycline	S	**R**	S
Streptomycin	S	S	**R**
Levomycetin	S	S	S
Kanamycin	S	S	S

Note: S – sensitive to antibiotics, R – resistant to antibiotics.

Growth for MPA$_{rich.}$ without added of antibiotics served as control cultures.

Among the studied mutant strains there were strains which changed sensitivity to antibiotics (tetracycline and streptomycin) (table 2).

To assess stability of the obtained mutant strains, they were re-inoculated during two months with two week interval to solid and liquid medium and the quantity of synthesized threonine was measured. The index of threonine synthesis from inoculation to inoculation was about the same (from 7,85 to 7,90 g/dm^3). By UV irradiation effect on threonine producers, *Brevibacterium flavum* IMV B-7446 mutant strain was obtained, which was different by more than 4 times from the stock culture in biosynthetic activity and amino acid production.

To identify the obtained mutant strain, analysis of its culture-morphological and physiological-and-biochemical properties was made. The auxotrophic mutant obtained after 24 hour growth on MPA is an aerobe in the form of gram-positive, asporogenic, often located at an angle bacilli measuring (0,7-1,0 x 2,0-4,0) μm. With the time the cells became shorter to coccus-like dimensions. During growth on MPA the colonies were round, bright, opaque, yellow in color (figure 6).

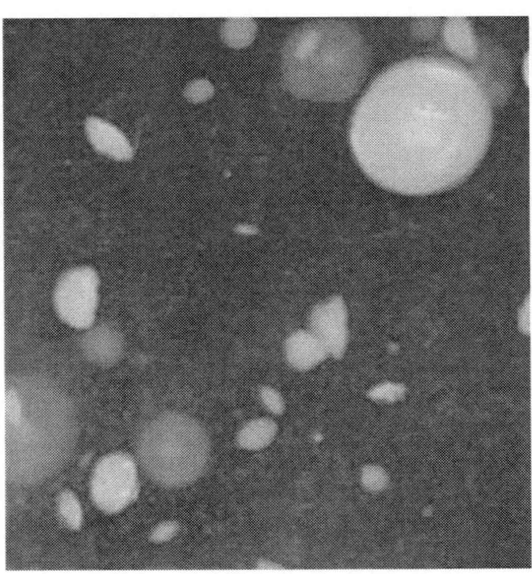

Figure 6. *Brevibacterium flavum* IMV B-7446 colonies in the MPA$_{rich.}$.

Cell growth occurred at temperature from 20 to 40°C (T$^0_{opt.}$ 31±1 °C), at pH from 4,0 to 9,0, (pH$_{opt.}$ 7,0). By analysis of culture-morphological and physiological-and-biochemical features the culture was identified as *Brevibacterium flavum* [36]. To confirm taxonomic position of the obtained mutant producer strain of threonine *Brevibacterium flavum* IMV B-7446, threonine producer strain 16S rRNA gene sequences were determined from the Collection and their phylogenetic position within the most related strains of *Brevibacterium* genus was established from «GenBank» database [32].

16S rRNA gene amplification was carried out using 27f and 1492r all-purpose bacterial primers with further sequencing.

To determine sibs the phylogenetic analysis was done and phylogenetic tree was built up. Considered as a measure of concordance for the set of aligned sequences of this topology is the measure (criterion) based on the principle of the highest similarity. Phylogenetic relationship tree diagram of *B. flavum* TH7, *B. flavum* IMB B-7446 strains was studied from the Collection of phylogenetically close representatives of *Brevibacterium* and *Corynebacterium* genera.

Below the phylogenetic tree is presented with the highest log-likelihood value – 2171,38, drawn up with the help of the method of maximum likelihood using Tamura-Nei model [34] for assessment of evolutionary distance. Number of bootstraps – 1000 (figure 8). The phylogenetic tree built by another statistical method - Neighbor-joining, had the same topology.

Comparative analysis of 16S rRNA gene nucleotide sequences with the sequences of this gene included in GenBank database showed that the investigated strains belonged to *Corynebacterium* genus and revealed the highest percentage of similarity (98%) to *C. glutamicum* species.

On the basis of the obtained results the tree diagram of phylogenetic relationships was built up between *B. flavum* TH7, *B. flavum* IMV B-7446 strains and several representatives of *Brevibacterium, Kocuria* and *Corynebacterium* genera (figure 7).

On the tree diagram the investigated strains are located in one group with the typical strains of *Corynebacterium glutamicum* species, this group is separated with 100% probability (bootstrap analysis) from another group formed by representatives of *Brevibacterium* and *Kocuria* genera. So the genetic analysis made showed that *B. flavum* TH7 and *B. flavum* IMV B-7446 strains belong to *C. glutamicum* species.

Numerous strain originally assorted to the genera *Brevibacterium, Corynebacterium, Microbacterium, Micrococcus* or *Artrobacter*, e.g., *Brevibacterium chang-fua, Brevibacterium divaricatum, Brevibacterium flavum, Brevibacterium glutamigenes, Brevibacterium lactofermentum, Brevibacterium roseum, Brevibacterium seonmiso, Brevibacterium sp.,*

Brevibacterium taipei, Brevibacterium thiogenitalis, Corynebacterium lilium, Corynebacterium herculis, Microbacterium ammoniaphilum, Microbacterium sp., Micrococcus maripunicenus, Artrobacter sp. wich have been included in numerical taxonomic studies were found to group into *C. glutamicum* clusters, indicating that they were misnamed [37].

Importantly, none of the amino acids-producing, nomenclatural *Brevibacterium* species turned out to represent a true member of the genus *Brevibacterium*. Consequently, numerous strains have been reclassified as *C. glutamicum*. Values reported in the literature for the genomic G+C content of *C. glutamicum* strains are in the range of about 53 to 58 mol %.

The precise G+C content of the genome sequence for *C. glutamicum* ATCC 13032 is 53,8 mol %, the precise G+C for mutant strain *B. flavum* IMV B-7446 is 55,4 mol%.

The discrepancy revealed in species identification based on complex of microbiological data and 16S rRNA gene sequencing analysis can be explained recently made re-classification of some *Brevibacterium* species as *Corynebacterium* ones [38]. Therefore, the more detailed analysis should be done for accurate identification of strains used in this chapter.

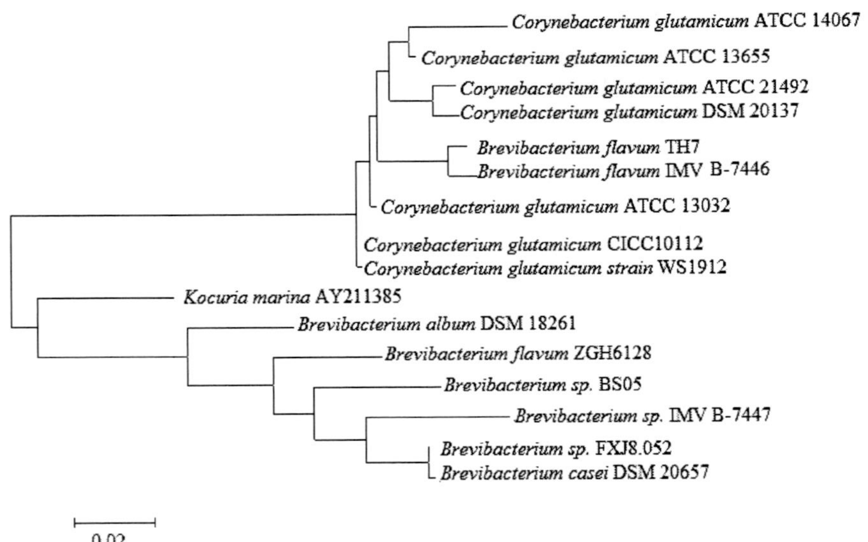

Figure 7. Phylogenetic position of *Brevibacterium flavum* TH7 and *Brevibacterium flavum* IMV B-7446 strains among representatives of *Brevibacterium, Kocuria* and *Corynebacterium* genera on the basis of 16S rRNA gene nucleotide sequences.

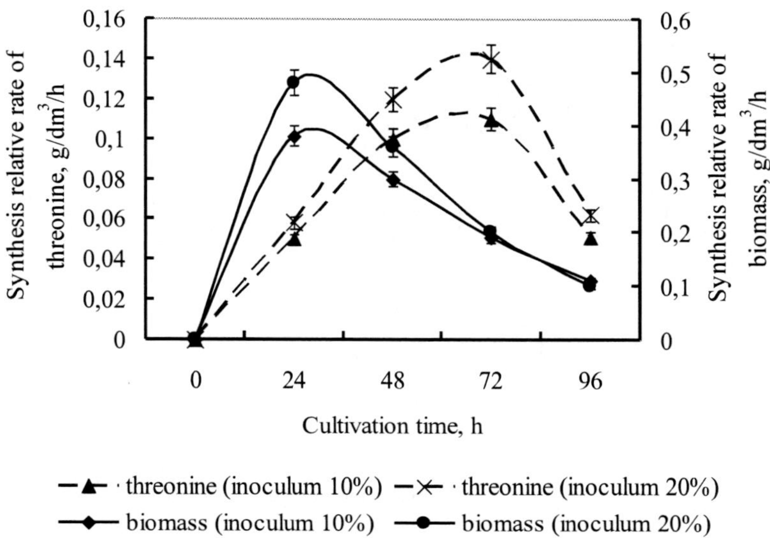

Figure 8. Synthesis rate of biomass and threonine.

2.4. Optimization of Mutant Strain Cultivating Conditions

Investigations in biotechnology of essential amino acids are aimed at set-up of optimal cultivating conditions, selection of commercially reasonable feedstock and selection of productive strains of the microorganisms which are capable of extracellular amino acid production. To provide higher accumulation of amino acids in the course of synthesis, the system of metabolic control has to be changed. To that end it is necessary to stimulate substrate consumption in some biosynthesis techniques as well as excretion of amino acids to medium, or to suppress adverse reactions and amino acid degradation processes [4, 7].

Under optimal cultivating conditions the speed of biomass accumulation and speed of waste product formation depend on inoculum quantity [4]. Influence of inoculum quantity on growth rate of bacterial population and threonine synthesis was investigated (figure 8). The best performance was observed when 20% of inoculate was introduced into medium.

One of the factors affecting microorganism growth and physiological activity is the temperature of microorganism cultivation. Accumulation of biomass and threonine synthesis against cultivation temperature is shown in figure 9.

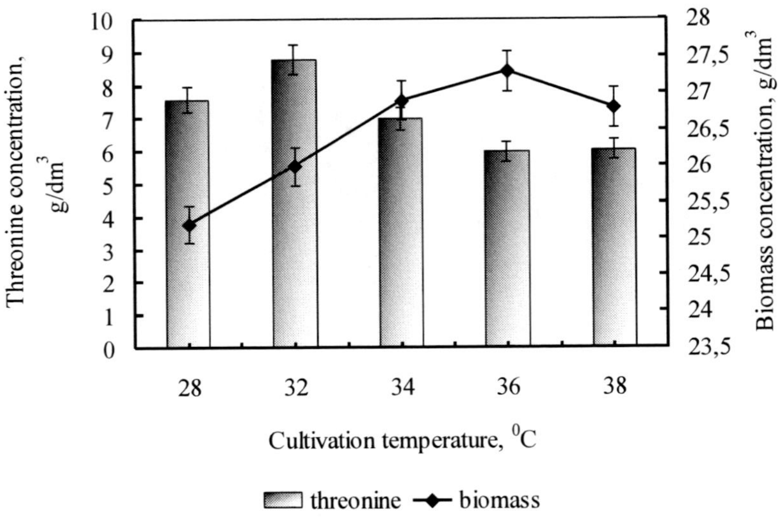

Figure 9. Temperature cultivation effect on the synthesis of biomass and threonine.

Temperature rise from 28 to 36^{0}C increased accumulation of biomass. Further temperature rise to 38^{0}C reduced biomass accumulation. Threonine synthesis, when temperature changed from 28 to 32 ^{0}C, rose to 8,8 g/dm^3, with further temperature increase synthesis dropped to 6,0 g/dm^3.

Further study was conducted at 32^{0}C. Accumulation of biomass and synthesis of threonine are significantly influenced by medium pH value [4]. For pH stabilization during threonine synthesis, Na_2CO_3 buffer agent was used. Effect of various Na_2CO_3 concentrations on pH, biomass growth and threonine synthesis was shown on figure 10. Maximal biomass accumulation occurred at Na_2CO_3 concentration of 0,3 g/dm^3 and pH 5,0. Maximal quantity of synthesized threonine was 9,3 g/dm^3. Increase in pH from 4,8 to 5,2 did not significantly affect biomass accumulation and threonine synthesis. Further cultivation of threonine producer was done at pH 5,0. Dynamics of biomass and threonine formation as a function of oxygen dissolution rate variation is given in figure 11.

The highest biomass accumulation occurred at oxygen dissolution rate of 10 g/O$_2$/dm^3/hour, and the biggest quantity of threonine accumulation – at 11 g/O$_2$/dm^3/hour. Higher dissolution speed suppressed growth of biomass and threonine. In further investigations oxygen dissolution speed of 11 g/O$_2$/dm^3/hour was used.

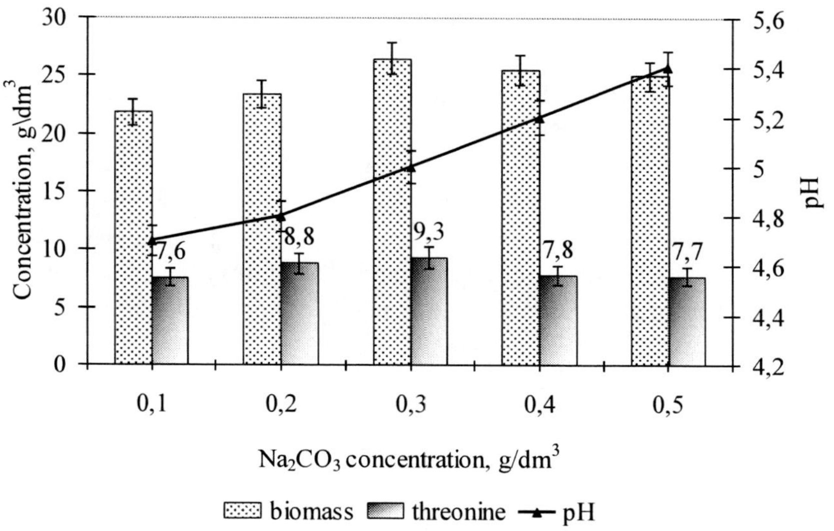

Figure 10. Effect of various Na_2CO_3 concentrations on pH, biomass growth and threonine synthesis.

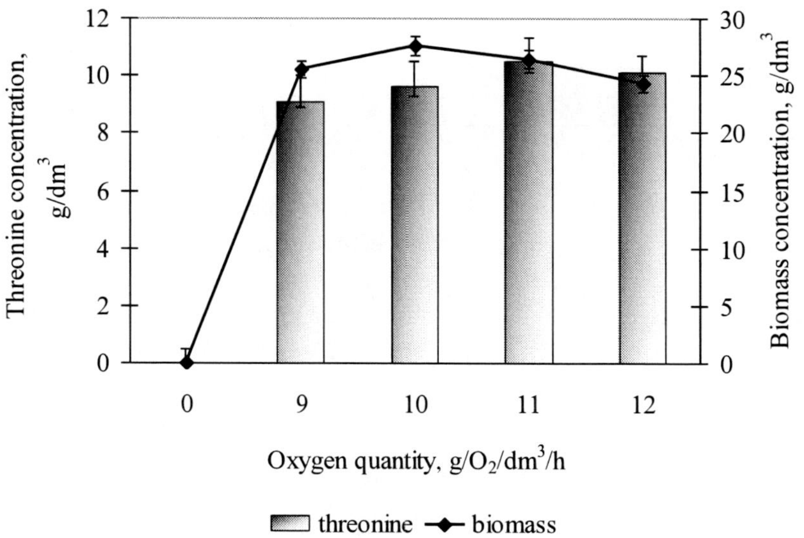

Figure 11. Influence of oxygen on the dissolution rate of biomass accumulation and threonine synthesis.

To optimize cultivation process the relationship between threonine concentration and various growth factors brought into the medium was studied. Their optimal concentrations (g/dm^3) were found: proline – 0,8; thiamin – 0,002; biotin – $2,0 \cdot 10^{-4}$; yeast extract – 1,0; mixture of isoleucine and methionine (0,4 g/dm^3 of each amino acid).

Effect of alternative carbon sources (beet molasses, milk serum, synthetic medium) on threonine synthesis was investigated. The best synthesis performance was with complex molasses medium compared to synthetic and serum one. The cultivation of the obtained mutant strain under conditions of the determined optimal parameters was conducted in molasses medium with addition of major mineral elements (table 3).

Table 3. The synthesis of amino acids and target conversion rate power supplies

Producers	Indicators culture fluid				
	pH	OD (1:10), λ_{440}	Concentration of sucrose, %	Concentration of threonine, g/dm^3	The conversion of sugar into the target amino acid, %
Starting medium	7,9±0,1	0,2±0,1	8,2±0,5	0,9±0,1	-
Threonine producers					
B. flavum TH 7	6,6±0,1	1,1±0,1	5,4±0,2	2,0±0,2	1,61±0,1
B. flavum IMV B-7446	6,7±0,1	1,4±0,1	1,4±0,3	11,9±0,3	16,2±0,2

Brevibacterium flavum TH7 stock strain had low level of threonine synthesis in molasses medium and consumed insignificant quantity of sugar (conversion – 1,61%) (table 3). *Brevibacterium flavum* IMV B-7446 mutant strain, when cultivated, produced 6 times more threonine in molasses mediums compared to the stock strain and consumed sugar intensively (conversion – 16,2 %).

CONCLUSION

1. UV irradiation effect was studied on vegetative cells of *B. flavum*, lethal (12 min.) and mutagenic (8 min.) irradiation dose was determined, and *Brevibacterium flavum* mutant strain was selected (as

to auxotrophy and resistance to β-oxynorvalin) with increased threonine productivity (7,9 g/dm^3).

2. Comparative analysis of 16S rDNA nucleotide sequences with the sequences of this gene included in GenBank database showed that the investigated strains belonged to *Corynebacterium* genus and revealed the highest percentage of similarity (98%) to *C. glutamicum* species, although additional analysis is necessary for more precise species definition

3. The growth medium was optimized for cultivation of producer and optimal cultivation parameters were determined (quantity of inoculum; growing temperature; aeration conditions). Under optimized cultivation conditions *Brevibacterium flavum* IMB B-7446 produced threonine in quantity of 11,9 g/dm^3.

ACKNOWLEDGMENTS

We would like to thank Lyubov Zelena for continuous support and discussion.

REFERENCES

[1]   Hermann, T. Industrial Production of amino acid by coryneform bacteria. *J. Biotechnol.*, 2003, v. 104, P. 155-172.

[2]   Tesseraud, S., Everaert, N., Boussaid-Om Ezzine, S., Collin, A., Metayer-Coustard, S. and Berri, C. Manipulating tissue metabolism by amino acids. *World's Poultry Sci. J.*, 2011, Vol. 67, P. 382-385.

[3]   Debabov V.G. *The threonine story*. Biochem. Engineer. Biotechnol., 2003, V. 79, P. 113-136.

[4]   Shulga, S.M., Tigunova, O.O., Tkachenko, A.F., Bejko, N.E., Andriiash, G.S., Pryyomov, S.G. Threonine biosynthetic process intensification. *Biotechnology*, 2011, V. 4, № 5, P. 97 – 102 (in Ukrainian).

[5]   Patent RU(11) 2 288 264(13) C2. Bacterium belonging to *Escherichia* genus as producer of L-threonine and method for preparing L-threonine / *Akhverdjan V., Savrasova E., Samsonova N., Ermishev V., Al'tman I., Ptitsyn L.* (RU) Date of publication: 27.11.2006, Bull. 33.

[6]   Dong, X., Quinn, P.J., Wang, X. Metabolic engineering of *Escherichia coli* and *Corynebacterium glutamicum* for the production of L-threonine. *Biotechnol. Adv.*, 2010, P. 1-13.

[7]   Andriiash, G.S., Zabolotna, G.M., Shulga, S.M. Regulation and intensification ways of lysine biosynthesis. *Microbiology and Biotechnology*, 2012, № 4, P. 6-17 (in Ukrainian).

[8]   Pometto, A., Shetty, K., Paliyath, G., Levin, R. E. *Food biotechnology*. Second Edition. United States: Taylor Francis Inc.; 2005.

[9]   Andriiash, G.S., Zabolotna, G.M., Shulga, S.M. Auxotrophity of producers of lysine. *Biotechnology*, 2012, V. 5, № 1, P. 70-77 (in Ukrainian).

[10]   Eikmanns, B. J., Metzger, M., Reinscheid, D., Kircher, M., Sahm, H. Amplification of three threonine biosynthesis genes in *Corynebacterium glutamicum* and its influence on carbon flux in different strains. *Appl. Microbiol. Biotechnol.*, 1991, V. 34, N 5, P. 617-622.

[11]   Verte`s, A.A., Inui, M., Yukawa, H. Manipulating *Corynebacteria*, from individual genes to chromosomes. *J. Appl. Environ. Microbiol.*, 2005, V. 71, № 12, P. 7633-7642.

[12]   Kalinowski, J., Bathe, B., Bartels, D., Bischoff, N., Bott, M., Burkovski, A. The complete *Corynebacterium glutamicum* genome sequence and its impact on the production of L-aspartate-derived amino acids and vitamins. *J Biotechnol.*, 2003, №104, P. 5–25.

[13]   Farfan, M.J., Aparicio, L., Calderon, I.L. Threonine overproduction in yeast strains carrying the *HOM3-R2* mutant allele under the control of different inducible promoters. *Appl. Environ. Microbiol.*, 1999, v.61, N 5, p. 110–116.

[14]   Komatsubara, S., Kisumi, M., Murata, K. Chibata, I. Threonine production by regulatory mutants of *Serratia marcescens*, *App. Env. Microbiol.*, 1978, P. 834-840.

[15]   Motoyama, H., Yano, H., Ishino, S., Anasava, H., Tesiba, S. Effect of amplification of the genes coding for the L- treonine biosynthetsc enzymes on the L- treonine production from methanol by gram-negative obligate methilotroph, *Methylobacillus glicogenes*. *Apl. Microbiol. Biotechnol.*, 1994, 42, P. 67-72.

[16]   Yang, H., Liao, Y., Wang, B., Lin, Y., Pan, L. Draft genome sequence of *Escherichia coli XH001*, a producer of L-threonine in industry. *J. Bacteriology*, 2011, 193(22), P. 6406-6407.

[17] Patec, M., Hochmannova, J., Nesvera, J. Production of Treonine by *Brevibacterium flavum* Contataining Treonine Biosynthethis Gene from *Escherichia coli. Folia Microbiol.*, 1993, 38 (5), P. 355-359.

[18] Guillouet, S., Rodal, A., Gorret, N., Lessard, P., Sinskey, A. Methabolic redirection of carbon flow toward isoleucine by expressing a catabolic threonine dehydratase in a threonine-overproducing *Corynebacterium glutamicum. Apl. Microbiol. Biotechnol.*, 2001, 57, P. 667-673.

[19] *Microbial Production of L-Amino Acids.* Advances in biochemical engeeneering biotechnology. V. 79. Managing editor: T. Scheper, Berlin Heidelberg: Springer-Verlag; 2003.

[20] Yetti, M.I., Pudjiraharti, S. Strain improvement of *Brevibacterium sp* ATCC 21866 for L-lysine production using ultra violet irradiation. *Technology Indonezia*, 2004, №27, P. 9-16.

[21] Andriiash, G.S., Zabolotna, G.M., Shulga, S.M. The mutant strains of microorganisms – producers of lysine and threonine. *Biotechnology Acta.* 2014, 3 (7), P. 95-101(in Ukrainian).

[22] *US Patent 2010/0159537 A1. Escherichia coli* strains that over-produce L-threonine and processes for their production. D'Elia J. N., Jordan S. W. Filed Jan. 24, 2010.

[23] Lee, J.H., Sung, B.H., Kim, M.S., et. al. Metabolic engineering of a reduced genome strain of *Escherichia coli* for L-threonine production. *Microbial Cell Factories*, 2009, 8, P. 1-12.

[24] Lee, H.-M., Lee, H.-W., Park, J.-H. Improved L-Threonine production of *Escherichia coli* mutant by optimization of culture conditions. *J. Biosci. Bioengineering*, 2006, V. 101, No 2, P. 127-130.

[25] Eikmanns, B.J., Metzger, M., Reinscheid, D., Kircher, M., Sahm, H. Amplification of three threonine biosynthesis genes in *Corynebacterium glutamicum* and its influence on carbon flux in different strains. *Appl. Microbiol. Biotechnol.*, 1991, V. 34, 5, P.617-622.

[26] Vagner de Alencar Arnaut de Toledo, Maria Claudia Colla Ruvolo-Takasusuki, Arildo José Braz de Oliveira, Emerson Dechechi Chambó and Sheila Mara Sanches Lopes. Spectrophotometry as a Tool for Dosage Sugars in Nectar of Crops Pollinated by Honeybees: Reducing sugar determination. In: Dr. Jamal Uddin (Ed.), *Macro to Nano Spectroscopy: Spectrophotometry as a Tool for Dosage Sugars in Nectar of Crops Pollinated by Honeybees.* InTech, 2012, P. 269-290.

[27] Wilson, K. Preparation of genomic DNA from bacteria. *Current protocol in Mol. Biol.* 2001, 2.4.1-2.4.5.

[28] Freidberg, E.C., Walker, G.C., Siede, W. *DNA repair and mutagenesis.* Washington: ASM Press; 1995.

[29] K. Edwards, J. Logan & N. Saunders, Eds. *Real-time PCR: An essential guide.* Wymondham, Norfolk, UK: Horizon Bioscience; 2004.

[30] Griffith, A.J.F., Miller, J.H., Suzuki, D.T., Lewontin, R.C., Gelbart, W.M. *An introduction to genetic analysis.* N.Y: W.H. Freeman; 2000.

[31] Guttel, R., Larsen, N., Woese, C. Lessons from an evolving rDNA:16S and 23S rDNA structures from a comparative perspective. *Microbiol. Review,* 1994, 8 (1), P.10-24.

[32] GenBank (Base sequences of DNA) URL: http://blast.ncbi.nlm.nih.gov.

[33] Thompson, JD, Higgins, DG, Gibson, TJ. CLUSTAL W: improving the sensitivity of progressive multiple sequence alignment through sequence weighting, position-specific gap penalties and weight matrix choice. *Nucl. Acids Res.,* 1994, 22(22), P. 4673–4680.

[34] Tamura, K, Nei, M. Estimation of the number of nucleotide substitutions in the control region of mitochondrial DNA in humans and chimpanzees. *Mol. Biol. Evol.,* 1993, 10, P. 512-526.

[35] Tamura, K., Stecher, G., Peterson, D., Filipski, A., Kumar, S. MEGA6: Molecular Evolutionary Genetics Analysis version 6.0. *Mol. Biol. Evol.,* 2013, 30, P. 2725-2729.

[36] *Bergey's Manual of Determinative Bacteriology. 9th Edition.* Edited by John G. Holt Baltimor: Williams & Wilkins;1994.

[37] Vertes, A., Inui, M., Yukawa, H. The biotechnological potential of *Corynebacterium glutamicum,* from *Umami* to *Chemurgy.* In: Yukawa H, Inui M.. *Corynebacterium glutamicum: Biology and Biotechnology: The biotechnological potential of Corynebacterium glutamicum, from Umami to Chemurgy.* Berlin Heldelberg: Springer-Verlag; 2013; 1-51.

[38] Liebl, W. *Corynebacterium* taxonomy. In: Eggeling E., Bott M. *Handbook of Corynebacterium glutamicum: Chapter 2. Corynebacterium taxonomy.* US, Boca Raton, Florida: CRC Press; 2005; 9-37.

In: Threonine
Editor: Jacob Coleman

ISBN: 978-1-63482-554-2
© 2015 Nova Science Publishers, Inc.

Chapter 2

THREONINE REQUIREMENTS FOR WHITE-EGG LAYERS

Matheus Ramalho de Lima[*,1],
Fernando Guilherme Perazzo Costa[2]
and Ricardo Romão Guerra[2]
[1]Federal University of the South of Bahia, Teixeira de Freitas,
Bahia, Brazil
[2]Federal University of Paraiba, Areia, Paraiba, Brazil

ABSTRACT

This study is aimed to determine the requirement of threonine for laying hens fed diets with increasing levels of threonine at variable and constant ratios. Two sequential experiments were conducted: the first wherein diets contained increasing levels of threonine, with variable digestible threonine: digestible lysine ratios; based on the results, the second experiment evaluated the requirement of digestible threonine associated with a constant amino acids: lysine ratio. In the first experiment, diets were formulated based on corn and soybean meal, supplemented with industrial amino acids L-lysine, DL-methionine, L-tryptophan, L-isoleucine and L-valine to meet nutritional requirements for laying hens, except for threonine. Evaluated digestible-threonine levels were 0.446, 0.486, 0.526, 0.565, 0.605 and 0.645%. In the second

---

[*] Corresponding author: mrlmatheus@gmail.com.

experiment, treatments consisted of diets formulated with corn and soybean meal, supplemented with amino acids L-lysine, DL-methionine, L-tryptophan, L-arginine, L-isoleucine and L-valine to meet the nutritional requirements for laying hens. The ratios between the amino acids and lysine were 91, 23, 90, 83 and 75% for digestible methionine + cystine, tryptophan, valine, isoleucine, and threonine, respectively. Performance and egg quality data were evaluated. Based on these data, it was determined that the threonine requirement of laying hens is 0.597 and 0.610%, or 684 and 626 mg/layer/day, with variable ratio and constant ratio at 75%, respectively.

Keywords: amino acid, egg quality, performance

INTRODUCTION

Threonine is an essential amino acid found at high concentrations in the heart, muscles, skeleton, and central nervous system. It is required for the formation of protein and maintenance of the body protein turnover, in addition to participating in the formation of collagen (Sá et al., 2007). Threonine is involved in other physiological functions such as digestion and immunity (Bisinoto et al., 2007). Mucus, a secretion produced by the gastrointestinal tract, is composed mainly of water (95%) and mucins (5%), which are glycoproteins with high molecular weight, especially rich in threonine. It is estimated that more than half of the consumed threonine is utilized in the intestine for the maintenance functions, used primarily in the mucin synthesis. The type and amount of mucin produced in the gastrointestinal tract affect the microbial communities (as it serves as substrate for bacterial fermentation and for fixation), the availability of nutrients (via endogenous loss of mucin and through the absorption of nutrients), and the immune function (by controlling the microbial population and nutrient availability) (Corzo et al., 2007).

Diets formulated with inadequate levels of digestible amino acids may lead to great losses concerning the effective production of these birds as well as economic and environmental factors. The ideal ratio between nutrients in poultry diets has long been sought, but the variable quality of foods, genetic lines, and raising specificities make it necessary that recommendations always be updated.

The advances in the knowledge of the protein metabolism and the advent of new synthetic amino acids, with large-scale commercial production at compatible prices, have allowed nutritionists to formulate diets closer to the animal requirements, improving the use of the dietary protein and reducing the costs and production of waste that is detrimental to the environment. Another advantage of using synthetic amino acids is the possibility of establishing an ideal ratio between all dietary amino acids by the concept of *ideal protein*, which contributes to the reduction of the dietary protein levels (Schimidt, 2009).

Given the abovementioned facts, the present study is aimed at determining the requirement of digestible threonine for white-egg layers fed diets containing increasing levels of digestible threonine with variable and constant threonine:lysine ratios.

THE STUDY AND DETAILS

The study was conducted in the Poultry Research Section of the Center for Agricultural Sciences of Universidade Federal da Paraíba, Brazil.

Two sequential experiments were carried out: the first with diets containing increasing levels of digestible threonine, wherein the digestible threonine: digestible lysine ratio was changed; and, based on the results of the former, a second experiment was conducted to evaluate the requirements of digestible threonine with a constant amino acids: lysine ratio.

Experiment 1 - Increasing Levels of Digestible Threonine, Variable Ratio

A total of 288 white-egg layers of the Dekalb White line at 29 weeks of age were distributed in a completely randomized design with six treatments, 12 replicates, and four birds per experimental unit. The experiment was divided into five 28-day phases, totaling 140 days of evaluation.

Diets were isoprotein and isocaloric (Table 1), consisting of corn and soybean meal, supplemented with industrial amino acids L-lysine, DL-methionine, L-tryptophan, L-isoleucine and L-valine so as to meet the nutritional requirement for white-egg layer hens in the laying phase, except for digestible threonine. The evaluated digestible-threonine levels were 0.446, 0.486, 0.526, 0.565, 0.605, and 0.645%. Crude protein was reduced to cause threonine deficiency, and the amino acids: lysine ratio met the recommendation of Rostagno et al. (2005), except for arginine, which was higher.

Table 1. Ingredients and chemical composition of the basal diet

Items, %	Threonine levels, %					
	0.446	0.486	0.526	0.565	0.605	0.645
Corn	67.675	67.675	67.675	67.675	67.675	67.675
Soy bean meal	18.011	18.011	18.011	18.011	18.011	18.011
Limestone	9.266	9.266	9.266	9.266	9.266	9.266
Dicalcium phosphate	1.548	1.548	1.548	1.548	1.548	1.548
Salt	0.486	0.486	0.486	0.486	0.486	0.486
L-lysine	0.257	0.257	0.257	0.257	0.257	0.257
DL-methionine	0.314	0.314	0.314	0.314	0.314	0.314
L-tryptophan	0.042	0.042	0.042	0.042	0.042	0.042
L-isoleucine	0.168	0.168	0.168	0.168	0.168	0.168
L-valine	0.143	0.143	0.143	0.143	0.143	0.143
Potassium carbonate	0.082	0.082	0.082	0.082	0.082	0.082
Choline	0.07	0.07	0.07	0.07	0.07	0.07
Vitamin and mineral mix[1]	0.10	0.10	0.10	0.10	0.10	0.10
Antioxidant[2]	0.01	0.01	0.01	0.01	0.01	0.01
Soybean oil	1.568	1.572	1.576	1.58	1.584	1.588
Inert[3]	0.000	0.009	0.018	0.027	0.036	0.045
L-glutamic acid	0.285	0.228	0.171	0.114	0.057	0.000
L-threonine	0.000	0.044	0.088	0.132	0.176	0.22
Total	100	100	100	100	100	100
Calculated composition						
Metabolizable energy (kcal/kg)	2900	2900	2900	2900	2900	2900
Crude protein (%)	14.5	14.5	14.5	14.5	14.5	14.5
Calcium (%)	4.02	4.02	4.02	4.02	4.02	4.02
Available phosphorus (%)	0.375	0.375	0.375	0.375	0.375	0.375
Sodium (%)	0.225	0.225	0.225	0.225	0.225	0.225
Chlorine (%)	0.353	0.353	0.353	0.353	0.353	0.353
Potassium (%)	0.58	0.58	0.58	0.58	0.58	0.58
Digest. lysine (%)	0.796	0.796	0.796	0.796	0.796	0.796
Digest. met + cys (%)	0.724	0.724	0.724	0.724	0.724	0.724
Digest. tryptophan (%)	0.183	0.183	0.183	0.183	0.183	0.183
Digest. valine (%)	0.716	0.716	0.716	0.716	0.716	0.716
Digest. isoleucine (%)	0.661	0.661	0.661	0.661	0.661	0.661
Digest. arginine (%)	0.838	0.838	0.838	0.838	0.838	0.838
Digest. threonine (%)	0.446	0.486	0.526	0.565	0.605	0.645
Thr:Lys ratio	56	61	66	71	76	81

[1]Mineral premix, per kg: Mn - 60 g; Fe - 80 g; Zn - 50 g; Cu - 10 g; Co - 2 g; I - 1 g; excipient q.s. - 500 g; vitamin premix (concentration/kg): vit. A - 15,000,000 IU; vit. D3 - 1,500,000 IU; vit. E - 15,000 IU; vit. B1 - 2.0 g; vit. B2 - 4.0 g; vit. B6 - 3.0 g; vit. B12 - 0.015 g; nicotinic acid - 25 g; pantothenic acid - 10 g; vit. K3 - 3.0 g; folic acid - 1.0 g; selenium - 250 mg; excipient q.s. - 1,000 g; and enough excipient for 1,000 g.

[2]Washed sand; [3]Ethoxyquin - 10 g.

Experiment 2 - Increasing Levels of Digestible Threonine, Constant Ratio

A total of 240 white-egg layers of the Dekalb White line at 29 weeks of age were distributed in a completely randomized design with five treatments, 12 replicates, and four birds per experimental unit. The experiment was divided into five 28-day phases, totaling 140 days of evaluation.

Treatments consisted of a diet formulated based on corn and soybean meal, supplemented with industrial amino acids L-lysine, DL-methionine, L-tryptophan, L-arginine, L-isoleucine and L-valine so as to meet the nutritional requirements for white-egg layer hens in the laying phase. Diets containing the same amino acids: lysine ratio, in accordance with the recommendations suggested by et al. (2005), except for digestible threonine: digestible lysine, which was kept constant at 75%. The ratios between amino acids and lysine were 91, 23, 90, 83 and 75% for digestible methionine + cystine, tryptophan, valine, isoleucine, and threonine, respectively, as shown in Table 2.

Analyzed Variables

Evaluated variables were: feed intake (g/layer/day); egg production (%); egg weight (g), egg mass (g/layer/day); conversion into egg mass (kg/kg) and dozen eggs (kg/dz); weight (g) and percentage (%) of yolk, albumen, and shell; shell thickness (mm); Haugh unit; and specific gravity.

The evaluation of egg production was divided into five 28-day periods. At the end of each period, orts from the diets from each plot were collected to calculate feed intake. Egg collection occurred twice daily (10h00 and 16h00), and laying frequency and mortality were recorded. Egg production in percentage was calculated by dividing the total number of eggs per plot by the number of birds, correcting mortality whenever it occurred. The eggs from the last three days of each period were weighed individually to obtain the average egg weight. Egg mass was calculated as the product between egg production and the average weight of the egg per plot. Feed conversion into egg mass was calculated as the ratio between feed intake and produced egg mass. Feed conversion into dozen eggs was calculated as the ratio between feed intake and production, and the result was multiplied by twelve. At the end of each period, four eggs were selected per plot to determine the weight and percentage of yolk, albumen, and shell, after manually separating these components and drying the shells were in an oven at 105 °C for four hours.

Table 2. Percentage and nutritional composition of the experimental diets

Ingredients, %	Digestible threonine levels (%)				
	0.507	**0.552**	**0.597**	**0.642**	**0.687**
Corn	67.807	67.563	67.318	67.071	66.832
Soybean meal	18.232	18.269	18.306	18.344	18.381
Limestone	9.309	9.308	9.308	9.307	9.307
Dicalcium phosphate	1.556	1.557	1.558	1.559	1.559
Soybean oil	2.031	1.96	1.889	1.814	1.746
Salt	0.523	0.524	0.524	0.524	0.524
DL-methionine	0.193	0.248	0.303	0.359	0.413
L-lysine	0.071	0.13	0.19	0.249	0.309
L-threonine	0.038	0.083	0.128	0.173	0.218
L-isoleucine	0.033	0.083	0.133	0.19	0.232
L-valine	0.017	0.071	0.125	0.179	0.234
L-tryptophan	0.01	0.024	0.038	0.051	0.065
Vitamin-mineral premix[1]	0.10	0.10	0.10	0.10	0.10
Choline	0.07	0.07	0.07	0.07	0.07
Antioxidant[2]	0.01	0.01	0.01	0.01	0.01
Total	100	100	100	100	100
Calculated composition					
Crude protein, %	14.130	14.349	14.567	14.792	15.005
calcium, %	4.02	4.02	4.02	4.02	4.02
Available phosphorus, %	0.375	0.375	0.375	0.375	0.375
Metabolizable energy, kcal/kg	2900	2900	2900	2900	2900
Digestible isoleucine, %	0.561	0.611	0.661	0.718	0.76
Digestible lysine, %	0.676	0.736	0.796	0.856	0.916
Digestible methionine + cystine, %	0.615	0.67	0.724	0.78	0.834
Digestible threonine, %	0.507	0.552	0.597	0.642	0.687
Digestible tryptophan, %	0.155	0.169	0.183	0.197	0.211
Digestible valine, %	0.608	0.662	0.716	0.77	0.824
Sodium, %	0.225	0.225	0.225	0.225	0.225
Chlorine, %	0.355	0.355	0.355	0.355	0.355
Potassium, %	0.580	0.580	0.580	0.580	0.580

[1]Mineral premix, per kg: Mn - 60 g; Fe - 80 g; Zn - 50 g; Cu - 10 g; Co - 2 g; I - 1 g; excipient q.s. - 500 g; vitamin premix (concentration/kg): vit. A - 15,000,000 IU; vit. D3 - 1,500,000 IU; vit. E - 15,000 IU; vit. B1 - 2.0 g; vit. B2 - 4.0 g; vit. B6 - 3.0 g; vit. B12 - 0.015 g; nicotinic acid - 25 g; pantothenic acid - 10 g; vit. K3 - 3.0 g; folic acid - 1.0 g; selenium - 250 mg; excipient q.s. - 1,000 g; and enough excipient for 1,000 g.

[2]Ethoxyquin - 10 g.

The percentage of each one of the egg components was obtained by dividing the weight of that component by the egg weight and then multiplying the result by 100. Shell thickness was measured with an electronic micrometer with 0.1 mm precision in three points on the egg midline, with which the arithmetic mean was calculated. At the end of each experimental period, representative samples of two eggs per plot were selected, and the eggs were immersed in different saline solutions with the proper adjustments for a volume of 25 L of water with densities varying from 1.060 to 1.100 with an interval of 0.0025 g/cm^3. Eggs were placed in the buckets with the solutions from the lowest to the highest density, and were collected when they floated. Respective values of the densities corresponding to the solutions of the containers were recorded. Before each evaluation, densities were checked with an oil densimeter.

For the histological analyses, biological fragments of the digestive (duodenum and liver) and reproductive (magnum and uterus) systems of 10 animals were collected randomly. Fragments were immersed in methacarn fixative solution (% methanol, 30% chloroform, and 10% acetic acid) for 12 h and then transferred to 70% alcohol.

For the optical microscopy, fragments were included in paraplast and subsequently sectioned into series with 5 μm in thickness. The following histological staining methods were performed: hematoxylin and eosin, Periodic acid-Schiff (PAS), and Masson's trichrome. Photomicrographs were captured with a micro-camera coupled to an Olympus BX-51 microscope, and images were digitalized on the KS 400.3 (Zeiss) software.

Results were subjected to analyses of variance and regression using the SAEG software (Universidade Federal de Viçosa, 2000).

RESULTS

Experiment 1 - Increasing Levels of Digestible Threonine, Variable Ratio

For the performance variables, only feed intake and conversion into dozen eggs were not influenced (P>0.05) by the treatments; however, the other variables underwent changes in these terms, providing results with which we can confirm the importance of digestible threonine in the performance of white-egg layers during the laying phase, since the main variables related to these birds, concerning production performance, were influenced statistically, e.g., egg weight, mass, and conversion into egg mass; experimental results regarding production performance are shown in Table 3.

Daily egg production was influenced (P<0.05) by the levels of threonine relative to lysine. With a quadratic response, production improved up to the ratio of 76% (0.605%), and according to the regression, the optimal determined ratio was 75.43%, which corresponds to 0.600% digestible threonine. For egg weight, mass, and conversion into mass, the response was similar to that observed for production, wherein the equations (displayed in Table 4), after derivations, determined the best ratios of 72.73, 73.70 and 72.80% digestible threonine: digestible lysine, or 0.579, 0.586 and 0.5795% digestible threonine, respectively.

Table 3. Feed intake (FI, g/layer/day), egg production (EP, %/layer/day), egg weight (EW, g/egg), egg mass (EM, g/layer/day), conversion into mass (CMS, kg/kg), and conversion into dozen eggs (CDZ, kg/12) according to the digestible threonine:lysine ratios

Threonine, %	FI	EP	EW	EM	CMS	CDZ
0.446 (56)	114.92	90.20	63.42	57.18	2.01	1.53
0.486 (61)	114.67	91.49	64.08	58.65	1.96	1.51
0.526 (66)	114.87	93.31	64.45	60.14	1.91	1.48
0.565 (71)	114.20	93.83	64.97	60.97	1.87	1.49
0.605 (76)	114.31	94.89	65.34	63.06	1.88	1.45
0.645 (81)	114.59	93.67	64.32	59.88	1.92	1.47
Mean	114.593	92.898	64.430	59.980	1.925	1.488
P value						
Linear	0.088	0.058	0.103	0.099	0.102	0.081
Quadratic	0.094	0.034	0.044	0.038	0.003	0.093
CV, %	1.578	2.075	1.51	2.541	3.052	2.354
SEM	0.0348	0.2061	0.0806	0.2397	0.0063	0.0034

Table 4. Polynomial equations of the variables influenced statistically by the digestible threonine:lysine ratios

Variable	Polynomial equation	R^2	Thr:Lys, %
Egg production	$\hat{Y} = -0.0112x^2 + 1.6896x + 30.352$	0.95	75.43
Egg weight	$\hat{Y} = -0.006x^2 + 0.8728x + 33.256$	0.86	72.73
Egg mass	$\hat{Y} = -0.0149x^2 + 2.1964x - 19.562$	0.80	73.70
Conversion into egg mass	$\hat{Y} = 0.0005x^2 - 0.0728x + 4.5332$	0.96	72.80

Table 5. Specific gravity (SG, g/cm^3), Haugh unit (HU), relative weights of albumen (ALBW, %), shell (SHLW, %) and yolk (YLKW, %), and shell thickness (SHLT, mm) of eggs according to the digestible threonine: lysine ratios

Threonine, %	SG	HU	ALBW	SHLW	YLKW	SHLT
0.446 (56)	1.085	96.16	58.75	14.06	27.13	0.4094
0.486 (61)	1.085	96.32	57.39	13.95	28.57	0.4074
0.526 (66)	1.086	96.38	59.08	14.36	26.58	0.4112
0.565 (71)	1.086	93.23	58.95	13.56	27.38	0.4003
0.605 (76)	1.085	96.85	58.9	13.22	27.84	0.4014
0.645 (81)	1.085	96.06	59.16	13.37	27.40	0.4048
Mean	1.0853	95.833	58.705	13.753	27.483	0.4057
P-value						
Linear	0.089	0.332	0.069	0.126	0.091	0.32
Quadratic	0.104	0.381	0.088	0.329	0.063	0.247
CV, %	0.183	2.132	3.071	6.793	7.542	1.034
SEM	0.00005	0.155882	0.078866	0.052632	0.080472	0.000522

As can be seen in Table 5, the variation in the digestible threonine: digestible lysine ratios did not result in any regression effect (P>0.05) regarding the egg-quality related variables.

Experiment 2 - Increasing Levels of Digestible Threonine, Constant Ratio

With a diet containing increasing levels of all amino acids, but with an always-constant ratio with lysine, the effects were similar to those observed in the previous experiment. Egg production, weight, mass, conversion into mass, and conversion into dozen eggs were significantly influenced by the treatments, with a quadratic response (Table 6).

Table 7 displays the polynomials originating from the statistical analyses of the variables presented in Table 6.

The internal and external egg quality was not influenced (Table 8) by the evaluated treatments, except for percentage of yolk, which had a quadratic response.

Table 9 shows the polynomial equation for the percentage of yolk in the eggs from the white-egg layers.

Table 6. Feed intake (FI, g/layer/day), egg production (EP, %/layer/day), egg weight (EW, g/egg), egg mass (EM, g/layer/day), conversion into mass (CMS, kg/kg), and conversion into dozen eggs (CDZ, kg/12) according to the digestible threonine levels

Thr:Lys	Thr, %	FI	EP	EW	EM	CMS	CDZ
	0.507	102.64	95.45	62.79	59.90	1.71	1.29
	0.552	103.29	97.32	63.86	62.15	1.66	1.27
75	0.597	102.31	97.50	64.07	62.46	1.64	1.26
	0.642	102.34	97.66	64.27	62.77	1.63	1.26
	0.687	102.73	95.79	63.81	61.12	1.68	1.29
Mean		102.662	96.744	63.76	61.68	1.664	1.274
P-value							
Linear		0.212	0.065	0.076	0.132	0.099	0.103
Quadratic		0.179	0.032	0.021	0.022	0.012	0.032
CV, %		1.55	1.65	2.25	2.74	3.27	2.51
SEM		0.0473	0.1243	0.0684	0.1401	0.0038	0.0018

Table 7. Polynomial equations of the variables influenced by the digestible threonine levels

Variable	Polynomial equation	R^2	Thr, %
Egg production	$\hat{Y} = -265.1x^2 + 318.8x + 1.979$	0.95	0.601
Egg weight	$\hat{Y} = -108.5x^2 + 135.1x + 22.24$	0.97	0.623
Egg mass	$\hat{Y} = -274.7x^2 + 334.8x - 39.2$	0.96	0.609
Conversion into egg mass	$\hat{Y} = 7.785x^2 - 9.504x + 4.533$	0.97	0.610
Conversion into dozen eggs	$\hat{Y} = 3.727x^2 - 4.504x + 2.619$	0.91	0.604

The weights of the layers from the different treatments with increasing concentrations of threonine with constant threonine: lysine ratio were not significant, although the weight of the group with 0.642% was lower (Figure 1).

The intestine of the layers on the treatment with 0.642% had longer villi, as also observed with the treatment with 0.687% (Figure 2A, B and C). Additionally, the villi of the hens from the treatment with 0.642% threonine showed a larger number of goblet cells, with higher levels of activity (Figure 2D, E and F).

The treatment with 0.687% threonine resulted in villi with a greater number of lamina propria in its interior (region where blood vessels are located) (Figure 3).

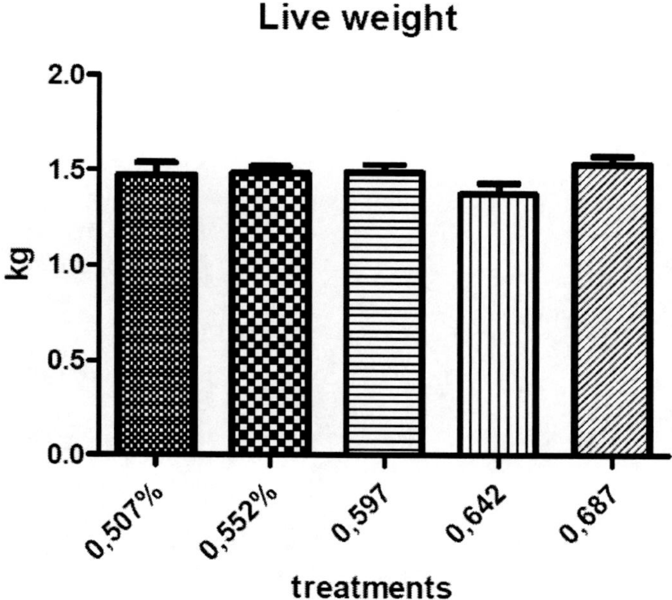

Figure 1. Live weight of the white-egg layers on treatments with increasing concentrations of threonine and a constant ratio with lysine.

Table 8. Specific gravity (SG g/cm^3), Haugh unit (HU), relative weights of albumen (ALBW %), shell (SHLW %) and yolk (YLKW %) and shell thickness (SHLT mm) of eggs according to the digestible threonine levels

Thr: Lys	Thr, %	SG	HU	ALBW	SHLW	YLKW	SHLT
	0.507	1.0864	97.16	57.81	15.67	26.52	0,4094
	0.552	1.0865	98.32	58.10	14.38	27.52	0,4074
75	0.597	1.0874	98.38	57.38	14.72	27.91	0,4112
	0.642	1.0882	97.23	57.00	15.21	27.79	0,4003
	0.687	1.0865	97.85	58.14	14.11	27.75	0,4014
Mean		1.087	97.788	57.686	14.818	27.498	0.40594
P-value							
Linear		0.335	0.367	0.056	0.146	0.19	0.365
Quadratic		0.534	0.349	0.456	0.367	0.045	0.098
CV, %		0.17	2.53	1.97	12.73	8.43	1.24
SEM		0.00004	0.069258	0.058466	0.075173	0.067493	0.00058

Table 9. Polynomial equations of the variables influenced statistically by the digestible-threonine levels

Variable	Polynomial equation	R^2	Threonine, %
Percentage of yolk	$\hat{Y} = -91.34x^2 + 115.1x - 8.294$	0.96	0.630

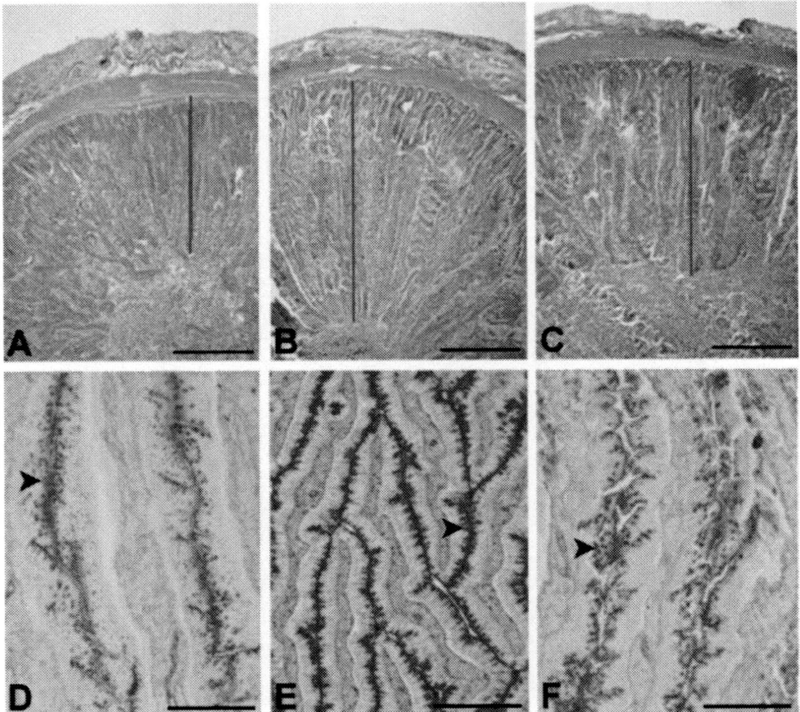

Figure 2. Photomicrographs of the intestine of layers treated with diets containing increasing concentrations of threonine, with a constant ratio with lysine. A and D - Intestine representing the treatments with 0.507, 0.552 and 0.597% threonine in the diet; B and E - Intestine representing the treatments with 0.642% threonine in the diet; C and F - Intestine representing the treatments with 0.687% threonine in the diet. Note that the intestinal villi in the treatment with 0.642% threonine are longer (vertical red lines) and have goblet cells (arrowheads) at a larger number and more active than in the other treatments. A, B and C - Hematoxylin-eosin stain (bar: 100 μm); D, E and F - Periodic acid-Schiff stain (bar: 400 μm).

The liver of the animals from the treatment with 0.687% were the only ones to show steatosis (lipid cytoplasmic hepatic vacuoles) (Figure 4C). This treatment provided greater deposition of collagen around the hepatic blood vessels (Figure 4F).

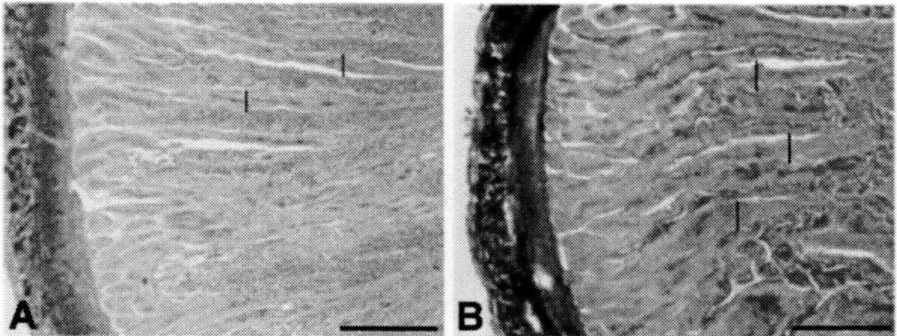

Figure 3. Photomicrographs of the intestine of layers treated with diets containing increasing concentrations of threonine, with a constant ratio with lysine. A - Intestine from the treatment with 0.507%; B - Intestine from the treatment with 0.687%. Note that the lamina propria (vertical lines) of the intestinal villi of the layers from the treatment with 0.687% is thicker. Masson's trichrome stain. Bar: 400 µm.

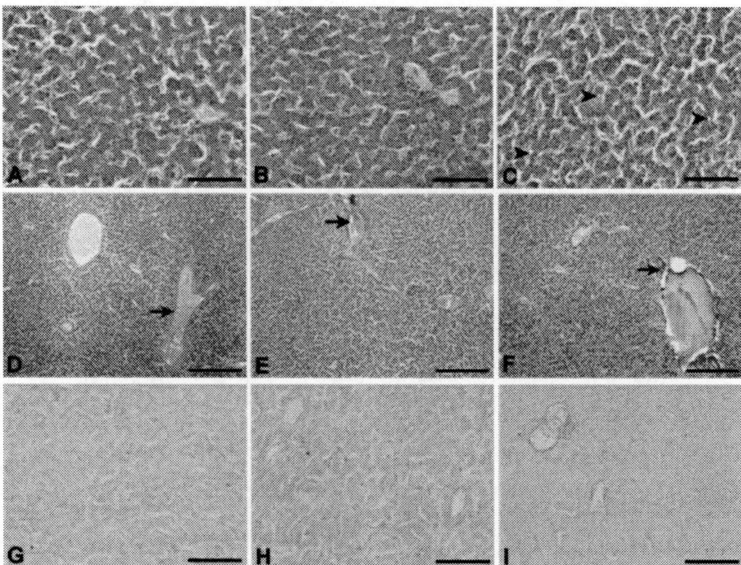

Figure 4. Photomicrographs of the liver of layers treated with diets containing increasing concentrations of threonine, with a constant ratio with lysine. A, D and G - Livers representing the treatments with 0.507, 0.552 and 0.597% threonine; B, E and H - Livers of animals from the treatment with 0.642% threonine; C, F and I - Livers of animals from the treatment with 0.687% threonine. Note the cytoplasmic lipid deposits (arrowheads) in the liver of the animals from the treatment with 0.687% threonine, as well as greater deposition of collagen in the hepatic blood vessels (arrows). Also note that the animals from the treatment with 0.642% threonine show a greater deposition of hepatic glycogen ("H") than the other groups. A, B and C - Hematoxylin-eosin staining; C, D and E - Masson's trichrome stain; F, G and H - Periodic acid-Schiff stain; A, B, C, G and H (Bar: 200 µm). D, E and F (Bar: 300 µm).

The liver of the animals from the treatment with 0.642% threonine, in turn, displayed greater deposition of glycogen (Periodic acid-Schiff stain) (Figure 4H).

With regard to the magnum, the animals from the treatment with 0.687% threonine showed a larger number of secondary folds as compared with the other studied treatments (Figure 5B).

The magnum of the animals on the treatment with 0.507% threonine, however, displayed only primary folds in many occasions (Figure 5A).

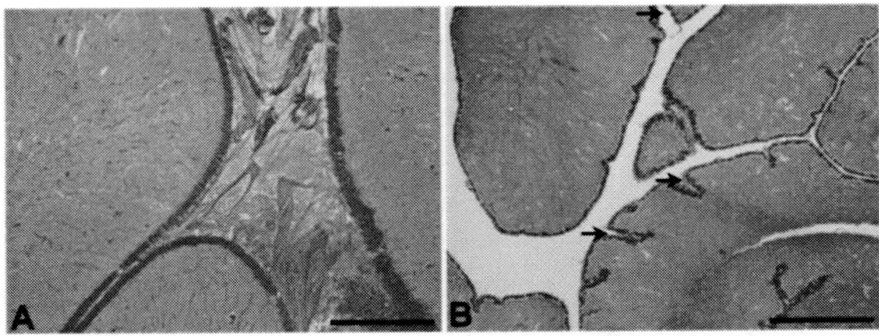

Figure 5. Photomicrographs of the magnum of layers fed diets with different percentages of threonine, keeping the ratio with lysine constant.

A - Magnum from the treatment with 0.507% threonine; B - Magnum from the treatment with 0.687% threonine.

Note that the number of secondary folds (arrows) of the magnum increases as the percentage of threonine in the diet is increased.

Periodic acid-Schiff stain. Bar: 1 mm.

DISCUSSION AND COMMENTS

Experiment 1 - Increasing Levels of Digestible Threonine, Variable Ratio

Evaluating the possible influences on the nutritional requirement of threonine in poultry, Samad et al. (2006) noted that genotype, sex, age, and efficiency of use of the dietary threonine provided a variation in the recommendations of this amino acid, except when the variable adopted for such recommendation was protein deposition, which, according to the authors, is not able to modify the dietary threonine requirement. According to the authors' statements, the constant re-evaluation of the nutritional recommendations for layer hens based on the characteristics mentioned as capable of modifying the nutritional requirements is remarkable. This evaluation is feasible given the availability of layer hen lines with increasingly shorter cycles in the market.

In this context, Faria et al. (2002) evaluated the nutritional recommendation of threonine for white-egg layers fed diets based on corn and soybean meal using eight treatments with decreasing levels of threonine and other amino acids. In diet 1 (Positive Control), the threonine level was 0.530%, whereas diet 2 contained 0.50% threonine. Diet 3 had 0.480% threonine, and the other amino acids were at 90% of their content in diet 2. For the subsequent diets, the threonine values were 0.45, 0.42, 0.40, 0.37 and 0.35%, with amino acids at 85, 80, 75, 70 and 65% of diet 2, respectively. After analyzing the results, the authors found that egg production, weight, and mass were significantly higher when layers consumed greater levels of threonine. Besides, birds fed threonine levels lower than 0.40% of the diet had weight loss, wherein the greatest loss occurred with the diet containing 0.35% threonine. The results shown by the authors confirm what was presented in experiments 1 and 2, because as the threonine levels in the first and second experiment were increased, performed increased as well, especially for the variables described by the abovementioned authors.

Sá et al. (2007) observed a quadratic effect for egg production and mass, estimating requirements of 0.499 and 0.501% digestible threonine, respectively, with 0.706% digestible lysine. In the present study, the significant effects were the same, but the determinations of the nutritional requirements were different, since the best results for egg production and mass were observed with 0.600 and 0.586% digestible threonine, corresponding to a ratio of 75.43 and 73.70% threonine with lysine. The similar effects with different recommendations can be explained by a commonly disregarded fact of extreme importance: the evaluated birds' line. Sá et al. (2007) evaluated Lohmman hens, while Dekalb hens were studied in the current experiment.

Based on increasing digestible-threonine levels in the diets, the results of this study show that the evaluated egg-quality traits were not influenced by the digestible threonine: digestible lysine ratios. Likewise, Valério et al. (2000) used increasing levels of threonine in the diet and did not observe improvement in the eggs' Haugh unit. However, Sá et al. (2007) observed a quadratic effect on Haugh unit as the digestible threonine level in the diet was increased, suggesting the dietary level of 0.478% digestible threonine for white-egg layers at 34 to 50 weeks of age.

Like Sá et al. (2007), Teixeira et al. (2005) reported that digestible threonine affects the egg quality, because the specific gravity of the eggshells was significantly improved when bids received diets containing higher levels of digestible threonine, and recommended 0.530% digestible threonine for white-egg layers. In our study no significant effect was observed on specific gravity.

The results obtained in this study corroborate the recommendation made by Gomez & Angels (2009), who conducted a study with layers in the second laying cycle with different levels of digestible threonine and digestible methionine and recommended the optimal digestible threonine: digestible lysine ratio of 74%. Cupertino et al. (2010), however, evaluated the requirement of digestible threonine for white- and brown-egg layers during the second laying cycle fed isoprotein and isoenergetic diets and concluded that diets with 0.446 and 0.465% digestible threonine, corresponding to digestible threonine:digestible lysine ratios of 68 and 71 for white- and brown-egg layers, respectively, lead to maximum performance in these birds.

In another study evaluating threonine levels in layer diets, but with birds in the second laying cycle, Shmidt et al. (2010) utilized diets with increasing levels of digestible threonine, with ratios of 58, 63, 68, 73 and 78% with digestible lysine (which was fixed at 0.653%), recommended 0.460% digestible threonine, which is equivalent to a digestible threonine: digestible lysine level of 70% for birds to achieve the best egg-production rates. Other determinations were obtained, e.g., 0.432, 0.451, 0.467, 0.452 and 0.436% digestible threonine, which correspond to ratios of 66, 69, 72, 69 and 67 for egg weight, mass, conversion into dozen eggs and into mass, and Haugh unit, respectively. Thus, the authors concluded that the best level of digestible threonine for layers in the second laying cycle is 0.467%, which corresponds to a ratio of 72% with digestible lysine.

Based on the results presented in this study, supplementation of L-threonine shows to be efficient on the performance of white-egg layers, because conclusive results are obtained with regard to the production-performance variables. However, this is not true for the variables that evaluate egg quality. It is possible that divergences in the applied methodologies or the simple lack of a significant effect on these traits may be responsible for this fact.

The methodology employed in our study shows that it is not recommended to evaluate nutritional requirements based on the amino acid level only, and nor is it to consider only the amino acids: lysine ratio, because the level of an amino acid depends on its ratio with the others, not only with lysine. Thus, it is not sufficient to formulate diets based on amino acids: lysine ratios, but the level of each amino acid is paramount to initially meet the birds' nutritional requirements.

Experiment 2 - Increasing Levels of Digestible Threonine, Constant Ratio

The birds' weight did not differ among the treatments; however, the weight of the animals from the group with 0.642% threonine was overall lower, which may indicate a greater allocation of energy to egg production instead of weight gain. The liver of the animals from this treatment was also more positive to Periodic acid-Schiff stain, which marks glycoproteins (in pink), among which is glycogen (source of energy). This result would indicate that these animals produce more glycogen, which would be positive from the production point of view. Since these animals are not heavier than the others, it is expected that they designate this source of energy to another location such as the reproductive system (egg production), thus corroborating the production findings.

Another histological result found in the liver of the animals fed diets with a greater percentage of threonine, more specifically in the treatment with 0.687%, is the presence of cytoplasmic vacuoles of lipids in the hepatocytes of the livers of the animals from that treatment, denoting moderate steatosis. Such pathology does not pose a risk to the animal health, and may be caused by the larger synthesis of estrogen in the ovary to give support to the greater egg production (Bunchasak & Silapasorn, 2005), which contradicts our results. Lima et al. (2011), in a study with quail, also observed the presence of steatosis when the percentage of threonine in the diet was increased. The increased dietary threonine also led to increased deposition of collagen around the hepatic blood vessels; however, there was no parenchymal deposition, which would denote a fibrotic process resulting from the continuous processes of liver cell regeneration, with cell death (Fausto et al., 2003). In quail, however, such deposition of parenchymal collagen was observed when threonine was increased in the diet (Lima et al., 2011).

The greater production by the animals with higher percentages of threonine in their diet found in the production analyses can also be explained by the greater length of the intestinal villi, mainly in the animals from the treatment with 0.642% threonine. The increased villus length provides a larger area of contact with the ingested feed and consequently a larger area of absorption. This re-arrangement of the mucosa demonstrates the physiological need for broadening the absorptive surface of the small intestine with regard to a certain nutrient (Aptekmann et al., 2001), which, in this case, is threonine. These facts corroborate the results obtained by Gomide-Júnior et al. (2004), which demonstrate that the intestinal mucosa responds to exogenous agents through morphological changes in the height and number of intestinal villi, depth of the intestinal crypts, cell proliferation, and number of dead cells per epithelial loss.

The animals fed the diet with 0.642% threonine also showed goblet cells (cells producing mucus from the intestinal villi) at a larger number and in a more active state than the animals from the other groups. This outcome is observed through the greater positivity to Periodic acid-Schiff stain (PAS). Regarding the greater positivity to the application of PAS on the intestinal epithelium and the increased threonine in the diets, it is demonstrated that the goblet cells from the intestinal epithelium are producing a larger amount of mucus, in addition to being in a larger number. This characteristic allows for the bolus to pass through the intestine more easily, thus preventing constipations and protecting the intestinal mucosa from damages caused by feed-deprivation or pathogenic agents (Gomide-Júnior et al., 2004). Hence, the increased threonine ratio in the diet would also lead to positive effects to the animal health. These characteristics observed in the intestine with increased dietary threonine were also observed in an experiment with quail (Lima et al., 2011).

Another trait that would improve production would be the greater thickness of the lamina propria of the intestinal villi found in the animals from the treatments with a higher percentage of threonine in the diet, especially in the treatment with 0.687%, because these are the areas (lamina propria) where blood vessels that capture the nutrients absorbed by cells (enterocytes) of the epithelium of the intestinal villus are located (Junqueira & Carneiro, 2003). Therefore, a greater villus thickness could also be related to a greater ability to pass the nutrients from the intestinal lumen to the bloodstream.

Lastly, the histophysiological studies also showed morphological differences in the structure of the magnum folds. In the treatments with a higher percentage of threonine in the diet, precisely the treatment with 0.687%, the magnum showed a greater number of secondary folds than as compared with the other studied treatments. Studies with quail have also shown increased number of secondary folds in the magnum (Lima et al., 2011), although in that study mucus production by the epithelium of this organ also increased, which did not occur in the present study. The increased number of secondary folds in the magnum (portion of the oviduct that produces the egg white - albumen) enlarges the area of contact with the egg being formed, which provides a faster formation of egg white.

According to the histophysiological results, the increased threonine content in the diet (treatments with 0.642 and 0.687%) provided favorable morphological conditions to increase in the layers' production, corroborating the production results.

FINAL CONSIDERATIONS

The adequacy of the threonine level in diets for Dekalb White layers is essential for better efficiency of these birds with regard to both the productive the physiological performances proved by histology.

The effects of amino acids cannot be evaluated with performance data only, because these are influenced after clear physiological and histological alterations, which were well observed in this study.

Based on these comments and results, we recommend threonine levels of 0.597 and 0.610%, or 684 and 626 mg/layer/day, with variable ratio and constant ratio at 75%, respectively, for white-egg layers.

ACKNOWLEDGMENTS

Special thanks to the Group for Studies of Poultry Technologies (*Grupo de Estudos em Tecnologias Avícolas*), CNPq, Guaraves, Granja Planalto, and Ajinomoto do Brasil.

REFERENCES

Aletor, V.A.; Hamid, I.I.; Nieb, E.; Pfeffer, E. Low-protein amino acid-supplemented diets in broiler chickens: effects on performance, carcass characteristics, whole-body composition and efficiencies of nutrient utilization. *Journal of the Science of Food and Agriculture*, v.80, p.547-554, 2000.

Andrade, L. Desempenho e qualidade dos ovos de poedeiras no primeiro e segundo ciclo de produção alimentadas com diferentes níveis de proteína bruta e aminoácidos na ração. 2004. 52f. Dissertação-Universidade Federal de Goiás, Goiânia, 2004.

Bisinoto, K.S.; Berto, D.A.; Caldara, F.R. et al. Relação treonina:lisina para leitões de 6 a 11kg de peso vivo em rações formuladas com base no conceito de proteína ideal. *Ciência Rural*, v.37, n.6, p.1740-1745, 2007.

Corzo, A.; Kidd, A.T.; Dozier, W.A. et al. Dietary threonine needs for growth and immunity of broilers raised under different litter conditions. *Journal of Applied Poultry Research*, v.16, p. 574-582, 2007.

Cupertino, E.S.; Gomes, P.C.; Vargas Jr, J.G. et al. Níveis nutricionais de treonina digestível para poedeiras comerciais durante o segundo ciclo de postura. *Revista Brasileira de Zootecnia*, v.39, n.9, p.1993-1998, 2010.

Faria, D.E.; Harms, R.H. and Russel, G.B. Threonine requirement of commercial laying hens fed a corn soybean meal diet. 2002. *Poultry Science,* 81:809-814.

Gomez, S., Angeles, M. Effect of threonine and methionine levels in the diet of laying hens in the second cycle of production *J. APPL. POULT. RES.* 2009 18: 452-457.

Hamilton, R.M.G. Methods and factors that affect the measurement of egg shell *quality. Poultry Science*, v.61, p.2022-2039, 1982.

Huyghebaert, G.; Butler, E.A. Optimun threonine requirements of laying hens. British Poultry Science , v.32, p.575-582, 1991.

Jardim Filho, R. M.; Stringhini, J. H. Qualidade de ovos, parâmetros bioquímicos e sanguíneos e desenvolvimento do aparelho reprodutor de poedeiras comerciais Lohmann LSL alimentadas com níveis crescentes de lisina digestível. Acta Scientiarum. *Animal Sciences*, v. 30, n. 1, p. 25-31, 2008.

NRC-National Research Council. Nutrient requirement of poultry. 9. ed. Washington, D.C.: National Academy Press, 1994.

Rostagno, H.S.; Albino, L.F.T.; Donzele, J.L.; Gomes, P.C.; Oliveira, R.F.; Lopes, D.C.; Ferreira, A.S.; Barreto, S.L.T. Tabelas Brasileiras para Aves e Suínos: Composição de Alimentos e Exigências Nutricionais. 2ª ed. UFV/DZO, 2005, 186p

Sá, L.M.; Gomes, P.C.; Cecon, P.R. et al. Exigência nutricional de treonina digestível para galinhas poedeiras no período de 34 a 50 semanas de idade. *Revista Brasileira de Zootecnia*, v.36, n.6, p.1846-1853, 2007.

Samad; Liebert, F. Modeling of threonine requirements in fast-growing chickens, depending on age, sex, protein deposition, and dietary threonine efficiency. 2006. *Poultry Science*, 85:1961-1968.

Schmidt, M.; Gomes, P.C.; Rostagno, H.S.; Albino, L.F.T.; Nunes,R.V.; and Cupertino, E.S. Exigência nutricional de lisina digestível para poedeiras semipesadas no segundo ciclo de produção. *Revista Brasileira de Zootecnia.*, v.38, n.10, p.1956-1961, 2009.

Shmidt, M.; Gomes, P.C.; Rostagno, H.S. Exigências nutricionais de treonina digestível para poedeiras no segundo ciclo de produção. *Revista Brasileira de Zootecnia,* v.39, n.5, p.1099-1104, 2010.

Teixeira, E. N. M.; Silva, J. H. V.; Lima, M.R. Exigência de treonina digestível para poedeiras leves e semipesadas. *Revista Brasileira de Ciência Avícola,* v. 7, supl., p. 131-131, 2005.

Valério, S. R. et al. . Determinação da exigência nutricional de treonina para poedeiras leves e semipesadas. *Revista Brasileira de Zootecnia*, Viçosa, v. 29, n. 2, Apr. 2000.

In: Threonine   ISBN: 978-1-63482-554-2
Editor: Jacob Coleman   © 2015 Nova Science Publishers, Inc.

Chapter 3

FLUORESCENCE, UV-VIS, AND CD SPECTROSCOPIC STUDY ON DOCKING OF CHIRAL SALEN-TYPE ZN(II) COMPLEXES AND LYSOZYME AND HSA PROTEINS

Tomoko Hayashi and Takashiro Akitsu
Department of Chemistry, Faculty of Science,
Tokyo University of Science,
Shinjuku-ku Tokyo, Japan

ABSTRACT

Threonine, one of proteinogenic as well as essential amino acids, is classified as polar and uncharged residue, which may play an important role in intermolecular interactions between protein molecules and small ligand molecules. Lysozyme of egg white contains seven threonine residues in primary structure, while HSA (human serum albumin), a certain transporting proteins in blood, of about 66 kDa, has some binding sites for external metal ions or metal complexes. We have prepared four new chiral salen-type Zn(II) complexes (**cyclo Zn**, **propane Zn**, **binaphtyl Zn**, and **Zn ntndd**) and characterized ^{1}H-NMR, fluorescence, UV-vis, and CD spectra. Among these Zn(II) complexes, **cyclo Zn** and **Zn tndd** are appropriate for the use of fluorescence proves against proteins because of intense emission when they are excited by UV light.

Gradual spectral changes of UV-vis and CD spectra elucidated docking of these Zn(II) complexes toward lysozyme or HSA accompanying with deformation of secondary structures of proteins. Not only quenching fluorescence intensity by energy transfer but also Stren-Volmer analysis of fluorescence spectra suggested that the numbers of binding sites of **Zn tndd**-HSA complex are larger than **cyclo Zn**-HAS or **cyclo-Zn**-lysozyme complexes.

INTRODUCTION

Dehydration condensation reaction of aldehyde and amine yields Schiff base compounds having C=N moiety, especially metal complexes incorporating Schiff base ligands made of bidendate amine (typically ethylenediamine) is called "salen-type" complexes commonly. Because of facility of preparation and variety of coordination geometries, salen-type complexes are important in view of magnetism [1-8], photofunctions [9-12], redox or electron transfer reactions. One of the most characteristic features of them may be large space at 3,3'-position around metal ions, which can be designed by introducing bulky substituent groups in the ligands. Salen-type ligands adopt planar form with flexibility, which provide structural variation of *cis*-[MN$_2$O$_2$] coordination environment (for example addition of axial ligands) to form MOFs (metal-organic frameworks) of chain structures [13]. Moreover, antenna effect associated with fluorescence is expected because coordination sites of metal ions are near to π-conjugated system of the ligands. Depending on metal ions various function can emerge, for example magnetism due to unpaired electron and redox for Cu(II), fluorescence for Zn(II), and magnetism, fluorescence, and NIR emission for additional Ln(III) ions [14-17].

Metal complexes have been used for catalysis of chemical reactions for a long time. Salen-type complexes are also employed especially for asymmetric Diels-Alder reactions and redox reactions [18-20]. Recently their various functions are widely applied for batteries or energy materials, biochemical or medical purposes (DDS or artificial metalloproteins) by many researchers besides coordination chemists [21-24].

Biochemical functions of metal complexes can be applied as medicine or drugs, for example antibacterial reagents, antitumor antibiotics, antipyretics, and hypoglycemic drugs [25, 26]. Due to molecular designs and resulting magnetic and emission properties, they also play an important role as test reagents [27, 28], for example DNA makers, biosensors, and microanalysis. Furthermore, molecular dynamics simulation predicts docking of metal complexes of some NSAIDs (non-steroidal anti-inflammatory agents) to proteins around space of higher order structures with hydrogen bonds [29], though systematic understanding has not been established based on structural information such as crystal structures and substituents effects.

In this way, many previous researches focused on biochemical approaches, while correlation between structures and electronic states of metal complexes has not been investigated sufficiently. Herein, we prepared new chiral salen-type Zn(II) complexes as fluorescence proves and investigate their spectroscopic properties and binding behavior to proteins, namely lysozyme of egg and HSA (human serum albumin) containing threonine residues.

EXPERIMENTAL SECTION

Materials

Chemicals of the highest commercial grade available (solvents are from Kanto Chemical, organic compounds are from Tokyo Chemical Industry, proteins are from Wako and metal sources from Wako and Aldrich) were used as received without further purification. Ultra pure water used was made with a Millipore Simplicity UV.

Preparation of Cyclo Zn

To a solution of *o*-vanillin (0.305 g, 2.00 mmol) dissolved in methanol (60 mL), (*1R,2R*)-(-)-1,2-cyclohexanediamine (0.114 g, 1.00 mmol) was added and stirred at 313 K for 2 h to give yellow solution of ligand. Zinc(II) acetate tetrahydrate (0.2217 g, 1.00 mmol) was added to the resulting solution and stirring at 313 K for 3 h to give yellow solution of the complex. After cooling the solution, this yellow compound was filtered. Yield 0.3872 g (81.20 %). Anal. Found: C, 56.55; H, 5.70; N, 5.53 %. Calc. for $C_{23}H_{27}N_2O_5Zn$: C, 57.93; H, 5.71; N, 5.87 %. IR (KBr (cm^{-1})): 462 (w), 471(w), 492 (w), 500 (w), 532 (w), 723 (m), 859 (w), 958 (w), 975 (w), 1023 (w), 1083 (m),1169 (w), 1219 (s), 1241 (s), 1326 (m), 1354 (w),1395 (m), 1448 (s), 1471(s), 1543(m), 1602(m), 1631(s) (C=N), 2342 (w), 2364(w), 2854(w), 2932(m), 3052(w), 3254 (sh), 3446 (sh). UV-Vis (diffuse reflectance): peak/cm^{-1} (F(R$_d$)); 24500 (1.21), CD (KBr): peak /cm^{-1} (θ/mdeg); 24570 (-40.19), 28490 (8.74). UV-Vis (acetone): peak/nm (Abs); 275 (0.65), 360 (0.27), CD (acetone): peak/nm (nm); 265 (8.15), 280 (-24.89), 350 (5.76), 385 (-17.27). ^{1}H-NMR (CDCl$_3$) /ppm: 8.25 (*s*, 2H, CH=N), 6.79 (*d*, 2H, Ph-H), 6.77 (*d*, 2H, Ph-H), 6.73 (*dd*, 2H, Ph-H), 3.36 (*s*, 6H, α-H), 3.12 (*t*, 4H, α-H), 2.35(*tt*, 4H, β-H), 2.07(*tt*, 4H, β-H).

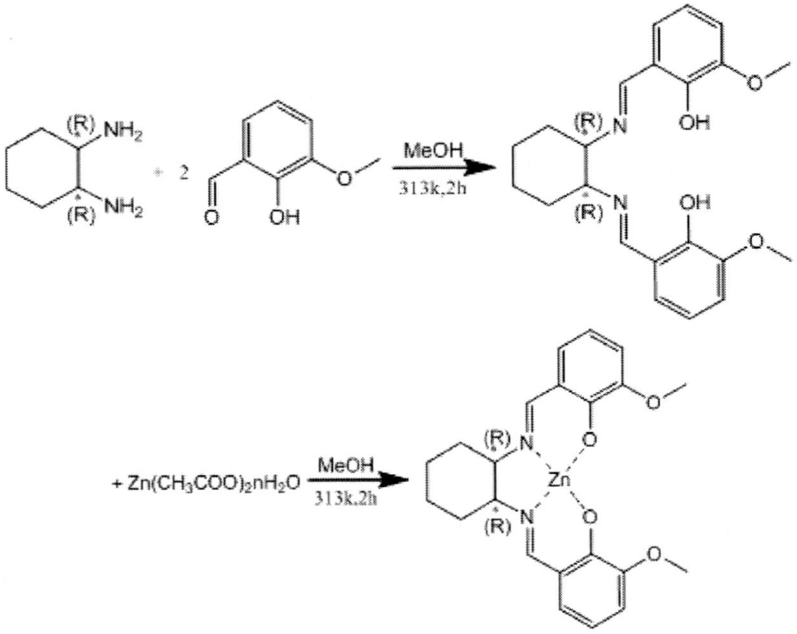

Scheme 1. Preparation of **cyclo Zn**.

Preparation of Propane Zn

To a solution of *o*-vanillin (0.305 g, 2.00 mmol) dissolved in methanol (60 mL), (*R*)-(+)-1,2-diaminopropanedihydrochloride (0.147 g, 1.00 mmol) was added and stirred at 313 K for 2 h to give yellow solution of ligand. Zinc(II) acetate tetrahydrate (0.2217 g, 1.00 mmol) was added to the resulting solution and the reaction was refluxed for 4 h at 373 K to give yellow solution of the complex. After cooling the solution, this yellow compound was filtered. Yield

0.1086 g (24.86 %). Anal. Found: C, 45.32; H,4.96; N, 6.31 %. Calc. for $C_{20}H_{23}N_2O_5Zn$: C, 55.12; H, 5.09; N, 6.43 %. IR (KBr (cm^{-1})): 464 (w), 472(w), 533(w), 612 (w), 738(s), 785 (m), 845 (m),969 (m), 1006 (w),1080 (w),1079 (m), 1098 (w), 1040 (w), 1171 (w), 1219 (s),1243 (s), 1286 (w), 1399 (w), 1449 (s), 1469 (s), 1499 (m), 1550 (w), 1636 (s) (C=N), 2353 (w),2358 (w), 2837 (w), 2932 (w), 2969 (w), 3435 (sh). UV-Vis (diffuse reflectance): peak/cm^{-1} (F(R_d)); 21880(3.95), CD (KBr): peak /cm^{-1} (θ/mdeg); 23200(-11.58), 29500(-3.85). UV-Vis (acetone): peak/nm (Abs); 270 (0.53), 349 (0.15), CD (acetone): peak/nm (nm); 265 (-1.22), 280 (3.56), 340 (-0.58), 380 (1.60). ^{1}H-NMR((CD_3)$_2$CO) /ppm: 8.53 (*s*,1H, CH=N), 8.46 (*s*, 1H, CH=N), 7.29 (*d*, 2H, Ph-H), 6.95(*d*, 2H, Ph-H), 6.75 (*dd*, 2H, Ph-H), 3.81(*sd*, 2H, α,β-H), 3.84 (*s*, 6H, α-H), 3.27 (*tq*, 1H, α-H),1.41(*d*, 3H, β-H).

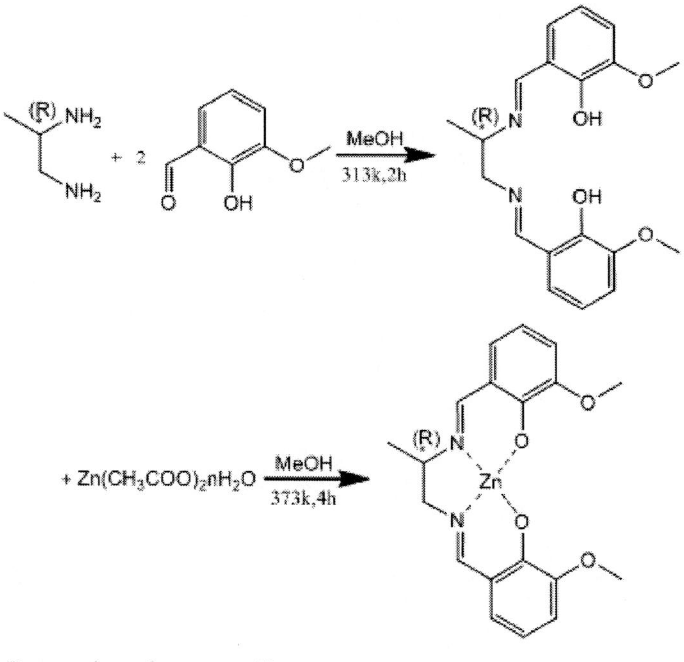

Scheme 2. Preparation of **propane Zn**.

Preparation of Binaphtyl Zn

To a solution of *o*-vanillin (0.305 g, 2.00 mmol) dissolved in methanol (60 mL), (*R*)-(+)-1,1'-binaphtyl-2,2'-diamine (0.284 g, 1.00   mmol) was added and stirred at 313 K for 2 h to give orange powder of ligand. Zinc(II) acetate tetrahydrate (0.2217 g, 1.00 mmol) was added to the resulting solution and the reaction was refluxed for 4 h at 373 K to give yellow compound of the complex. Yield 0.3279 g (35.75 %). Anal. Found: C, 76.01; H, 4.98; N, 4.97 %. Calc. for $C_{37}H_{29}N_2O_5Zn$: C, 68.68; H, 4.52; N, 4.33 %. IR (KBr (cm^{-1})): 480 (w), 493 (w), 500 (w), 504 (w), 532 (w), 567 (w), 578 (w), 619 (w), 632 (w), 688 (w), 716 (w),735 (m) ,748 (m),782 (s), 820 (m) , 840 (s), 864 (w), 971 (m), 1079 (w), 1083 (w), 1204 (m), 1252 (s), 1342 (w), 1402 (w), 1461 (s), 1503 (w), 1573 (w), 1590 (w), 1606 (s) (C=N), 2343 (w), 2358 (w), 2830 (w), 2931 (w), 2996 (w), 3055 (w), 3434 (sh). UV-Vis (diffuse reflectance): peak/cm^{-1} ($F(R_d)$)); 20750(3.22), 23750(2.99), CD (KBr): peak /cm^{-1} (θ/mdeg); 24570(-24.21). UV-Vis (acetone): peak/nm (Abs); 285 (0.98), 325 (0.36), 385 (0.26), CD (acetone): peak/nm (nm); 275 (19.52), 330 (-36.64), 375 (13.34), 415 (-10.46). ^{1}H-NMR(CDCl$_3$) /ppm: 7.98 (s, 2H, CH=N), 7.95 (s, 2H, Np-H), 7.90 (s, 2H, Np-H), 7.87(d, 4H, Np-H), 7.39 (dd, 4H, Np-H), 7.00 (d, 2H, Ph-H), 6.85(d, 2H, Ph-H), 6.55(t,2H,Ph-H), 3.93 (s,6H, α-H).

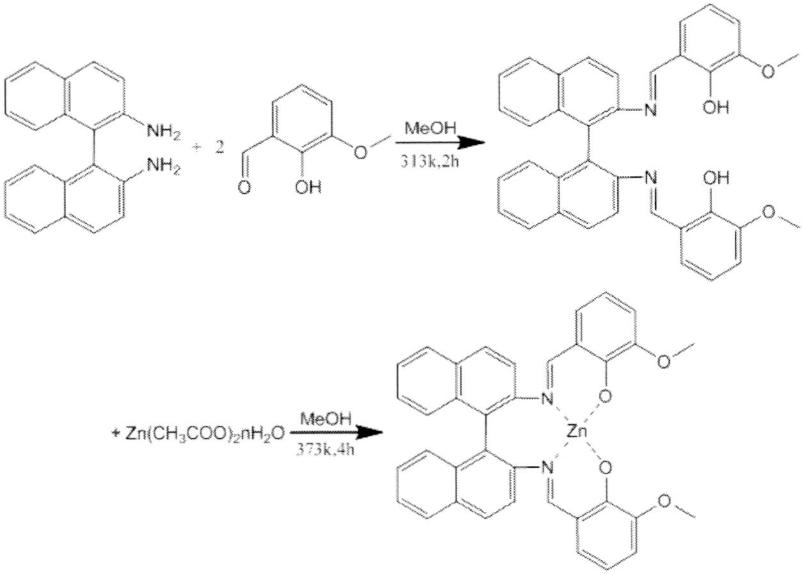

Scheme 3. Preparation of **binaphtyl Zn**.

Preparation of Zn tndd

To a solution of 2-hydroxy-5-methylisophtalaldehyde (0.328 g, 2.00 mmol) dissolved in acetonitrile (60 mL), (*1R,2R*)-(-)-1,2-cyclohexanediamine (0.114 g, 1.00mmol) was added and stirred at 313 K for 2 h to give yellow powder of ligand. Zinc(II) acetate tetrahydrate (0.2217 g, 1.00 mmol) was added to the resulting solution and stirring at 313 K for 24 h to give yellow powder of the complex. The solid product was filtered, washed with acetonitrile, and dried. This compound was dissolved in methanol (60 ml), then 1,3-diaminopropane was added and stirred at 313 K for 3h to give yellow compound of complex. Yield 0.3279 g (35.75 %). Anal. Found: C, 76.01; H, 4.98; N, 4.97 %. Calc. for $C_{37}H_{29}N_2O_5Zn$: C, 68.68; H, 4.52; N, 4.33 %. IR (KBr (cm^{-1})): 411 (w), 420 (w), 439 (w), 457 (w), 481 (w), 490 (w), 508 (w), 525 (w),565 (w), 579 (w), 669 (w), 752 (w), 779 (w), 804 (w), 830 (w), 870 (w), 984 (w), 1031 (w), 1093 (w), 1144 (w), 1168 (w), 1230 (m), 1261 (w), 1294 (w), 1318 (w),1338 (w), 1402 (m), 1453 (m), 1544 (m), 1619 (s) (C=N), 2359 (w), 2856 (m), 2925 (m), 3432 (sh). UV-Vis (diffuse reflectance): peak/cm^{-1} ($F(R_d)$); 24570 (2.32), CD (KBr): peak /cm^{-1} (θ/mdeg); 21930 (-1.17), 24330 (4.65), 27780 (-1.01), 29850 (-4.64). UV-Vis (acetone): peak/nm (Abs); 270 (0.76), 395 (0.57) CD (acetone): peak/nm (nm); 255 (115.09), 270 (23.07), 285 (-29.81), 380(-15.93), 405 (8.50), 425 (-6.36). ^{1}H-NMR(CDCl$_3$) /ppm: 8.01 (s,2H, CH=N), 7.64 (*s*, 2H, Ph-H), 6.83 (*s*, 2H, Ph-H), 3.75(*s*, 4H, β-H), 3.30(*t*, 4H, β-H), 3.28(*t*, 2H, β-H), 2.12 (*s*, 6H, α-H), 1.80 (*dt*, 4H, β-H), 1.60(*tt*, 4H, β–H), 1.57(*tt*, 2H, β-H).

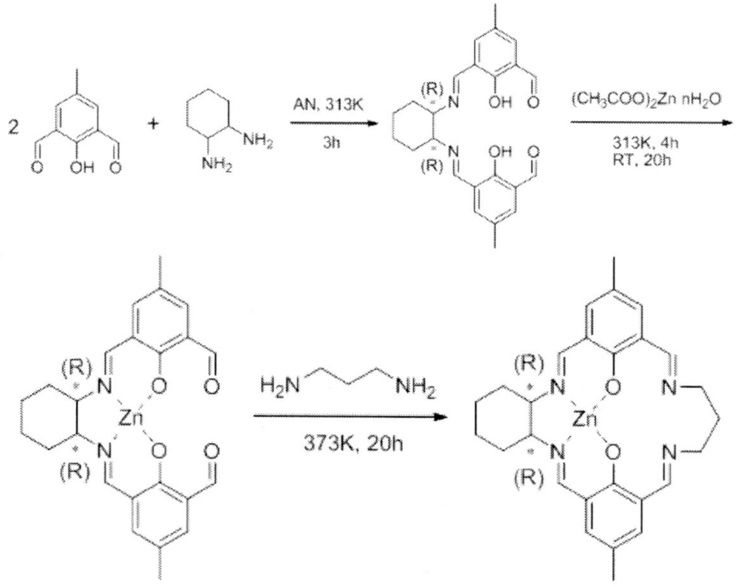

Scheme 4. Preparation of **Zn tndd**.

Physical Measurements

Elemental analyses (C, H, N) were carried out with a Perkin-Elmer 2400II CHNS/O analyzer at Tokyo University of Science. Infrared spectra were recorded as KBr pellets on a JASCO FT-IR 4200 plus spectrophotometer in the range of 4000-400 cm^{-1} at 298 K. Electronic spectra were measured on a JASCO V-570 UV/VIS/NIR spectrophotometer (equipped with an integrating sphere for diffuse reflectance spectra) in the range of 800-200 nm at 298 K. Circular dichroism (CD) spectra were measured as KBr pellets on a JASCO J-820 spectropolarimeter in the range of 800-200 nm at 298 K. Fluorescence spectra were recorded on a JASCO FP-6200 spectrophotometer at 298K. ^{1}H NMR spectra were recorded on a JEOL JMN-300 spectrometer (300 MHz) in CDCl$_3$ or (CD$_3$)$_2$CO.

Preparation of Solutions for Docking Studies

Lysozyme, from Egg White and albumin, from Human Serum (HSA) stock solutions, 10.0-80.0 μM, was prepared in sodium citrate buffer of pH 4.25, and then stored at 273–277 K in a refrigerator. **Cyclo Zn** stock solutions was prepared with the same method, and **Zn tndd** stock solutions were first prepared in methanol, and then diluted with sodium citrate buffer, 10.0-80.0 μM.

For UV-Vis spectroscopy, the concentration of the complex sample was constant (30 μM) while the protein stock solutions concentration was from 0 to 40.0 μM. For fluorescence spectroscopy, the concentration of the protein stock solutions was constant (30 μM) while the complex concentration was from 0 to 40.0 μM. Docking experiments of metal complexes-proteins were carried out by using mixture solutions of proteins and **cyclo Zn** and **Zn tndd** exhibiting strong emission. UV-vis spectra were measured under constant concentration of 30 μM for Zn(II) complexes on increasing protein concentration at 0, 5, 10, 15, 20, 25, 30, 35, 40 μM. As for the **Zn tndd**-HSA composite system, UV-vis spectra were measured under constant concentration of 30 μM for HSA on increasing **Zn tndd** concentration at 0, 5, 10, 15, 20, 25, 30, 35, 40 μM. Fluorescence spectra were measured under constant concentration of 30 μM for protein on increasing Zn(II) complexes concentration at 0, 5, 10, 15, 20, 25, 30, 35, 40 μM. Commonly biochemical docking experiments between small molecules and proteins or DNA were detected by using UV-vis spectra, fluorescence spectra, and computer simulations [25, 26, 30-32]. In this study, we employed two typical proteins indicating various biochemical functions (lysozyme [33, 34] and HAS [29, 35] showing emission) of different molecular weights. Stern-Volmer diagrams (relative intensity of proteins to intensity without quenchers (metal complexes) vs. concentration of added quenchers) [25, 26] were made by using intensity of fluorescence accompanying with quenching. Parameters, namely Stern-Volmer constant K_{SV}, binding constant $K\alpha$, and the numbers of binding sites in protein n, were evaluated according to equations (1) and (2), where F_0 and F denote fluorescence intensity without or with quenchers, respectively, and $[Q]$ denotes concentration of added quencher (metal complexes) [26].

$$\frac{F_0}{F} = 1 + K_{SV}[Q] \qquad \cdots (1)^{26}$$

$$\ln\frac{F_0 - F}{F} = \ln K_a + n\ln[Q] \qquad \cdots (2)^{26}$$

RESULTS AND DISCUSSION

Electronic and CD Spectra

Diffuse reflectance electronic spectra and CD (circular dichroism) spectra were measured for **cyclo Zn**, **propane Zn**, **binaphtyl Zn**, and **Zn tndd**, respectively. In solutions, electronic spectra and CD spectra were also measured for them (Table 1) and assignment was estimated and confirmed based on the results of TD-DFT calculation of **Zn tndd** (not shown). The present results are similar to the analogous complexes [36-45].

Fluorescence Spectra

Figure 1 shows fluorescence spectra of **cyclo Zn**, **propane Zn**, **binaphthyl Zn** and **Zn tndd**. Fluorescence spectra were recorded with an excitation wavelength at 280 nm, for lysozyme and HSA, while at 360 nm, for Zn complexes. For excitation at λ_{ex} = 360 nm (Figure 1 [above]), emission peaks appeared at λ_{em} = 491, 488, 523, and 471 nm for **cyclo Zn**, **propane Zn**, **binaphthyl Zn** and **Zn tndd**, respectively. For excitation at λ_{ex} = 280 nm

(Figure 1 [below]), emission peaks appeared at λ_{em} = 430-513 and 470 nm for **cyclo Zn** and **Zn tndd**. The fluorescence peak of **cyclo Zn** was broaden. These peaks were derived from MLCT of Zn(II) ion. The peak intensity of 360 nm is stronger than that of 280 nm.

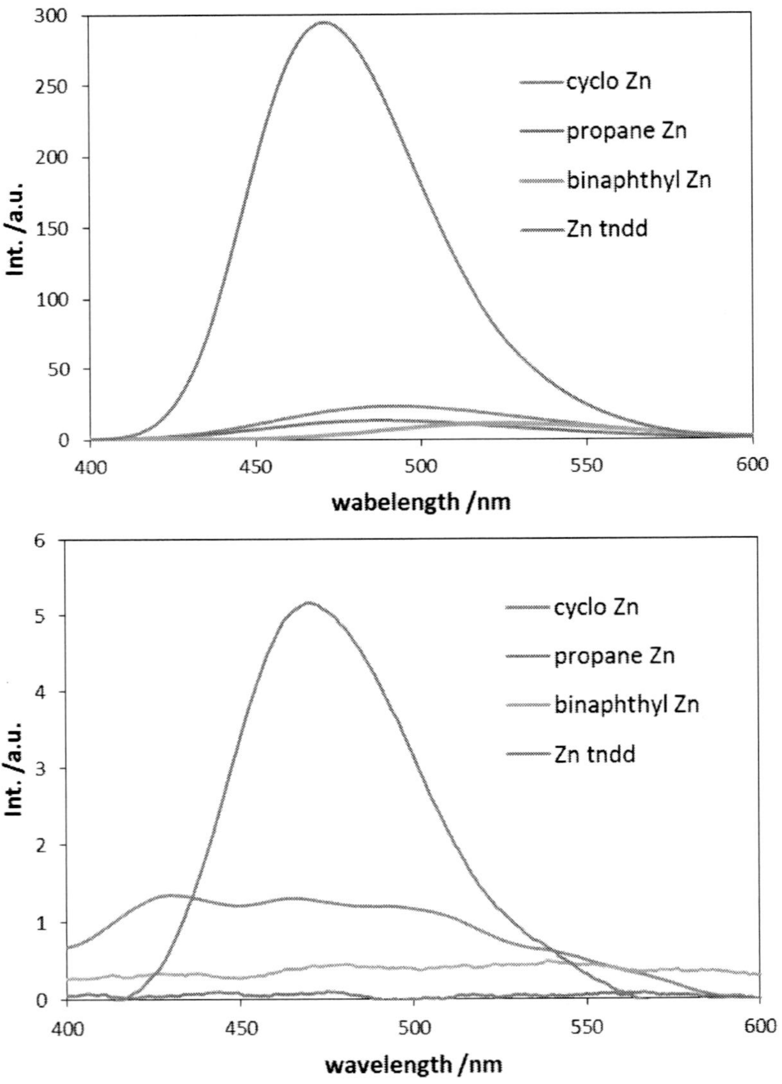

Figure 1. Fluorescence spectra in 30μM methanol solutions ([above] λ_{ex} = 360 nm; [below] λ_{ex} = 280 nm) for **cyclo Zn**, **propane Zn**, **binaphthyl Zn** and **Zn tndd**.

Table 1. Summary of UV-vis and CD spectra

	CD		UV-vis		assignment
	λ /nm	θ /mdeg	λ /nm	Abs.	
cyclo Zn	265	8.15	275	0.65	$\pi-\pi^*$
	280	-24.89			
	350	5.76	360	0.27	CT
	385	-17.27			
propane Zn	265	-1.22	270	0.53	$\pi-\pi^*$
	280	3.56			
	340	-0.58	349	0.15	CT
	380	1.6			
binaphthyl Zn	275	19.52	285	0.98	$\pi-\pi^*$
	330	-36.64	325	0.36	
	375	13.34	385	0.26	CT
	415	-10.46			
Zn tndd	255	115.09			
	270	23.07	270	0.76	$\pi-\pi^*$
	285	-29.81			
	380	-15.93			
	405	8.50	395	0.57	CT
	425	-6.36			

Among four chiral Zn(II) complexes, only the CD spectrum of **Zn tndd** exhibited opposite Cotton effect. Since intensity of (solid state) fluorescence spectra around 450-500 nm of **cyclo Zn** and **Zn tndd** was strong, **cyclo Zn** and **Zn tndd** are found to be promising candidate as fluorescence proves.

Study of Protein-Binding with UV Spectroscopy

Figure 2 shows UV-Vis absorption spectra of Zn complexes in the absence and presence of the proteins, in **cyclo Zn**-Lysozyme composite system, **cyclo Zn**-HSA composite system and **Zn tndd**-HSA composite system. In Figures 2 and 3, the absorbance of **cyclo Zn** increased with increasing concentration of lysozyme and HSA.

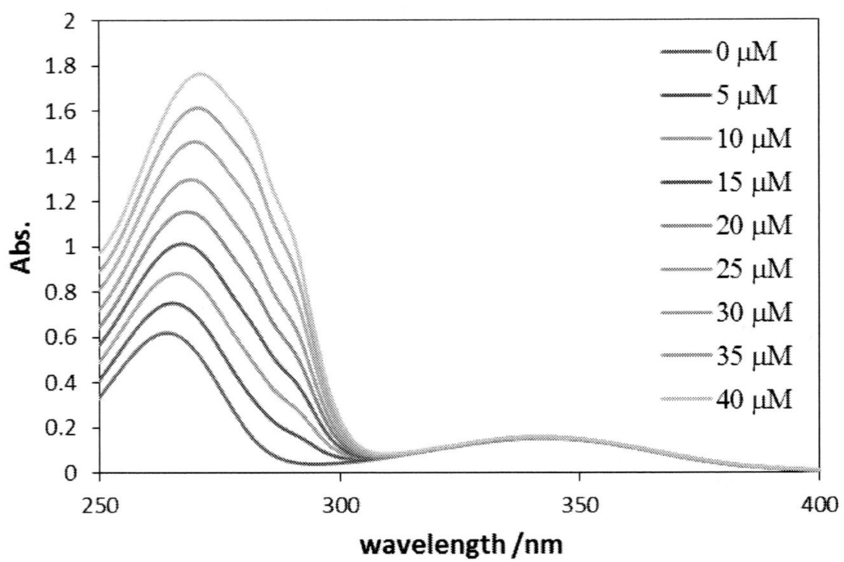

Figure 2. UV-Vis absorption spectra of **cyclo Zn**-Lysozyme composite system, [**cyclo Zn**] = 30 μM, [lysozyme] = 0, 5, 10, 15, 20, 25, 30, 35, 40 μM.

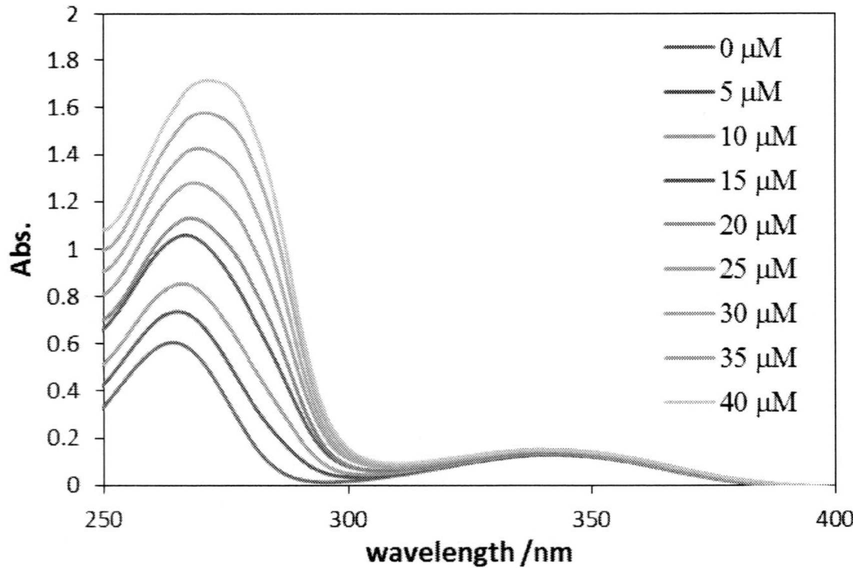

Figure 3. UV-Vis absorption spectra of **cyclo Zn**-HSA composite system, [**cyclo Zn**] = 30 μM, [HSA] = 0, 5, 10, 15, 20, 25, 30, 35, 40 μM.

Also, the maximum peak position around 280 nm shows a red shift. The peak around 350 nm due to charge transfer of **cyclo Zn** exhibited no change. The changes of UV-Vis absorbance, due to structural changes of Zn(II) complexes, indicate occurrence of the interaction between Zn(II) complexes and proteins [46, 47].

In Figure 4, the absorbance of **Zn tndd** increased with increasing concentration of lysozyme and HSA but there is no shift of maximum peak around 280 nm. Therefore, the red shift in Figures 2 and 3 is derived from $\pi-\pi*$ bands of **cyclo Zn**. It occurred due to changes in the microenvironment of π–conjugated of the complex because of docking with proteins. Additionally, lack of isosbestic points suggested not equilibrium of two components but new composites formed by docking of metal complexes and proteins.

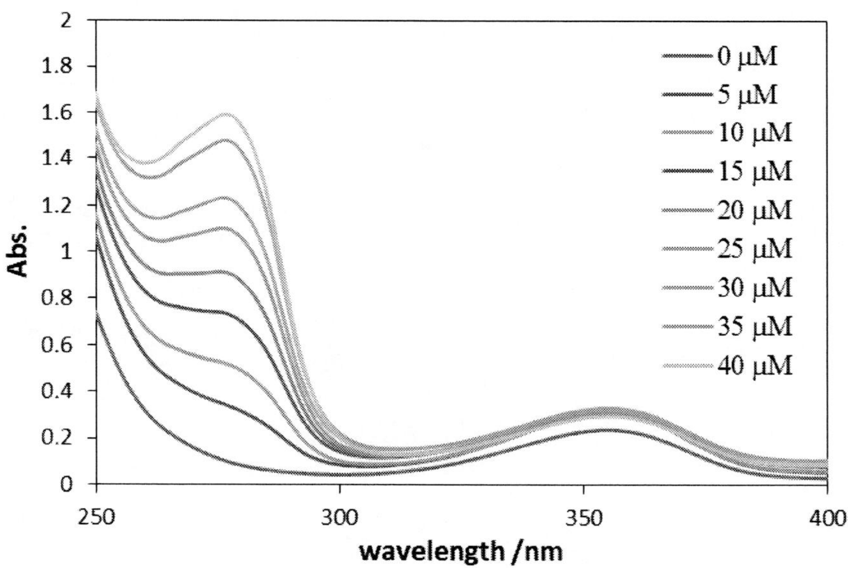

Figure 4. UV-Vis absorption spectra of **Zn tndd**-HSA composite system, [**Zn tndd**] = 30 µM, [HSA] = 0, 5, 10, 15, 20, 25, 30, 35, 40 µM.

Figure 5 shows UV-Vis absorption spectra of HSA in **Zn tndd**-HSA composite system, which is absence or presence of **Zn tndd**. The maximum peak position around 280 nm increased with increasing concentration **Zn tndd** and it shows a blue shift slightly. It indicates that **Zn tndd** forms a composite system with HSA and the absorption intensity increases to expose the molecule of a chromogenic peptide chain [46]. Figure 6 shows CD spectra of **Zn tndd**-HSA composite system, which is absence or presence of HSA.

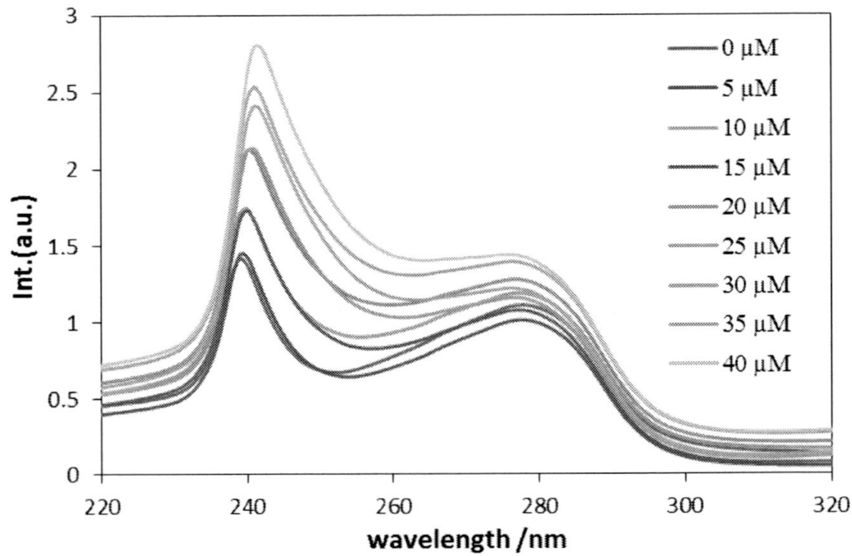

Figure 5. UV-Vis absorption spectra of **Zn tndd**-HSA composite system, [HSA] = 30 µM, [**Zn tndd**] = 0, 5, 10, 15, 20, 25, 30, 35, 40 µM.

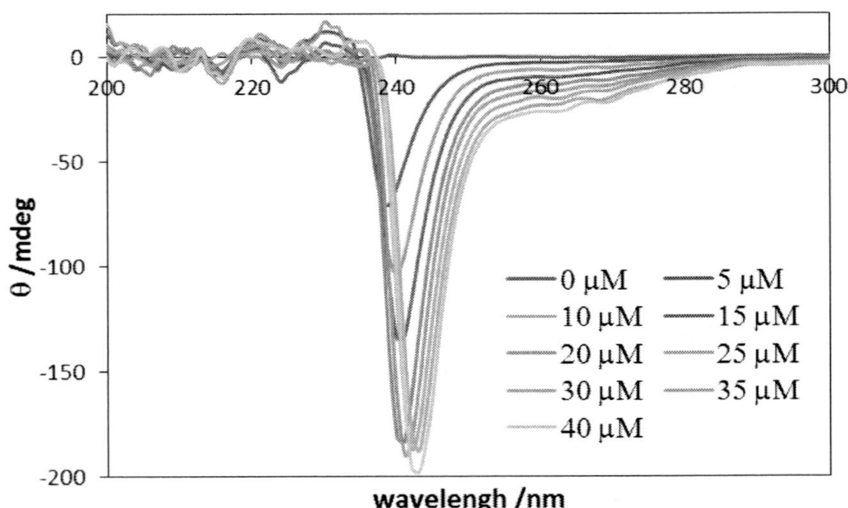

Figure 6. CD spectra of **Zn tndd**-HSA composite system, [**Zn tndd**] = 30 µM, [HSA] = 0, 5, 10, 15, 20, 25, 30, 35, 40 µM.

The negative peak around 240 nm is due to n-π* transfer of peptide bond in α–helix [46, 47]. The red shift of the peak supports the interaction between **Zn tndd** and HSA and conformational changes of HSA. Therefore, Zn complexes form composite system with proteins because of their interaction.

Study of Protein-Binding with Fluorescence Spectroscopy

Figures 7-9 depict fluorescence spectra of proteins at various concentrations of the Zn(II) complexes, namely **cyclo Zn**-Lysozyme composite systems, **cyclo Zn**-HSA composite systems, and **Zn tndd**-HSA composite systems. In Figures 7 and 8, the fluorescence of the tyrosine residues and the tryptophan residues of lysozyme and HSA, around 340 nm, decreased with increasing concentration of **cyclo Zn**.

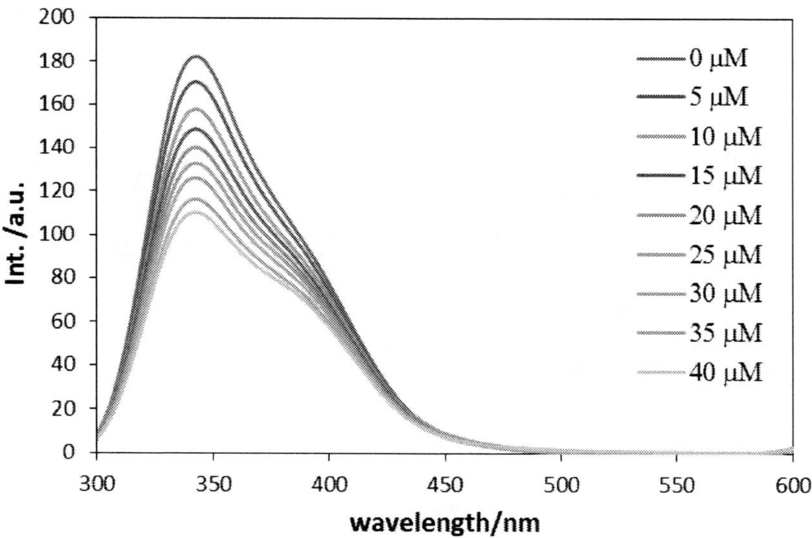

Figure 7. Fluorescence spectra of **cyclo Zn**-lysozyme composite system, [**cyclo Zn**] = 30 µM, [lysozyme] = 0, 5, 10, 15, 20, 25, 30, 35, 40 µM, λ_{ex}=295 nm.

So it has interaction between proteins and **cyclo Zn** to occur energy transfer between proteins and complexes. In Figure 9, the fluorescence of the tyrosine residues and the tryptophan residues of HSA, around 340 nm, decreased with increasing concentration of **Zn tndd** similar to the preceding fluorescence peak around 500 nm increased.

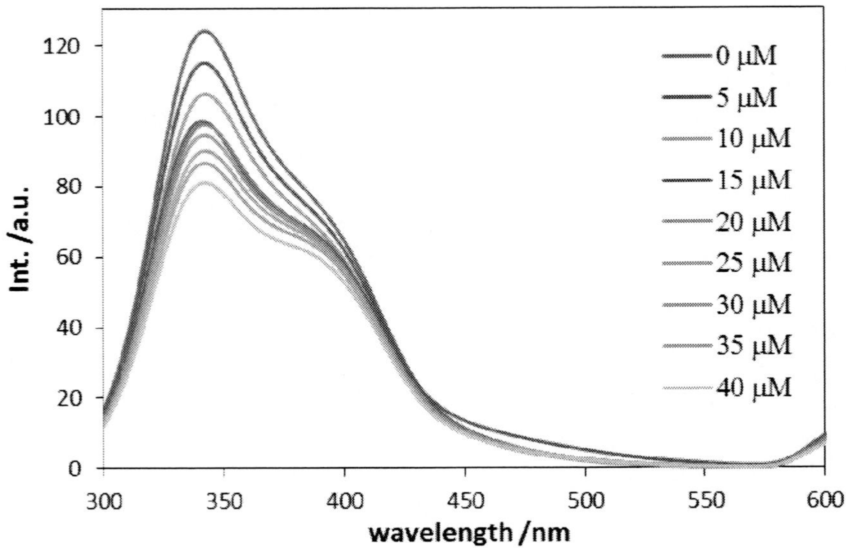

Figure 8. Fluorescence spectra of **cyclo Zn**-HSA composite system, [**cyclo Zn**] = 30 µM, [HSA] = 0, 5, 10, 15, 20, 25, 30, 35, 40 µM, λ_{ex}=280 nm.

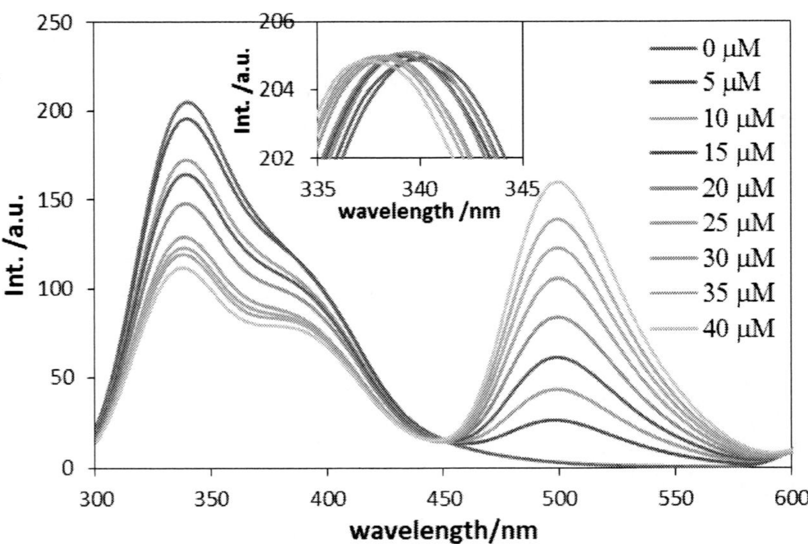

Figure 9. Fluorescence spectra of **Zn tndd**-HSA composite system, [**Zn tndd**] = 30 µM, [HSA] = 0, 5, 10, 15, 20, 25, 30, 35, 40 µM, λ_{ex}=280 nm.

According to fluorescence spectra in methanol solutions, this fluorescence peak is derived from **Zn tndd**. Though fluorescence should be weak in λ_{ex}=295 nm as it was, it shows strong peak in Figure 9. It supports that the interaction between **Zn tndd** and HSA is occurred and energy transfer from HSA to **Zn tndd**.

Inner spectra of Figure 9 are fluorescence spectra aligned the intensity of the peak around 340 nm due to HSA. The fluorescence peak shows blue shift slightly, which suggests change of polarity or hydrophobic in microenvironment around the tyrosine residues and the tryptophan residues and presence of interaction with Zn(II) complexes.

Energy loss of **Zn tndd**-HSA composite system is smallest of three composite systems. Furthermore in this system, we have succeeded in bringing out two fluorescence wavelengths, which is derived from the **Zn tndd** and the HSA, from one excitation wavelength.

Stern-Volmer Analysis

We made Stern-Volmer plots using the peak intensity at 340 nm of fluorescence spectra in order to determine the state of docking proteins with Zn(II) complexes. Figures 10-12 denote Stern-Volmer plots of **cyclo Zn**-Lysozyme composite system, **cyclo Zn**-HSA composite system, and **Zn tndd**-HSA composite system, respectively. In the above figures, a vertical axis is F_0/F (a.u.) and a horizontal axis is the concentration of Zn(II) complexes (mol L^{-1}). The calibration curve, whose y denotes intensity ratio and x denotes concentration of Zn(II) complexes had the positive slope. K_{SV} was obtained from this slope of the calibration curves.

The below figures were made by taken the logarithm of both axes of each upper plots. The calibration carve had the positive slope too, and $K\alpha$ and n were calculated. Table 2 shows a summary of parameters, which are K_{SV}, $K\alpha$ and n of each composite system.

As elucidated by previous studies [25, 26, 33-35], the numbers of the binding site n was approximately 1. It indicates that proteins are docking with Zn(II) complex at the rate of 1:1. Especially, the number of the binding site is larger than the other two in **Zn tndd**-HSA composite system. In **Zn tndd**-HSA composite system, Stern-Volmer constant K_{SV} and binding constant K_α are the largest in three composite systems. Stern-Volmer constant is an indicator of efficiency of the quenching constant.

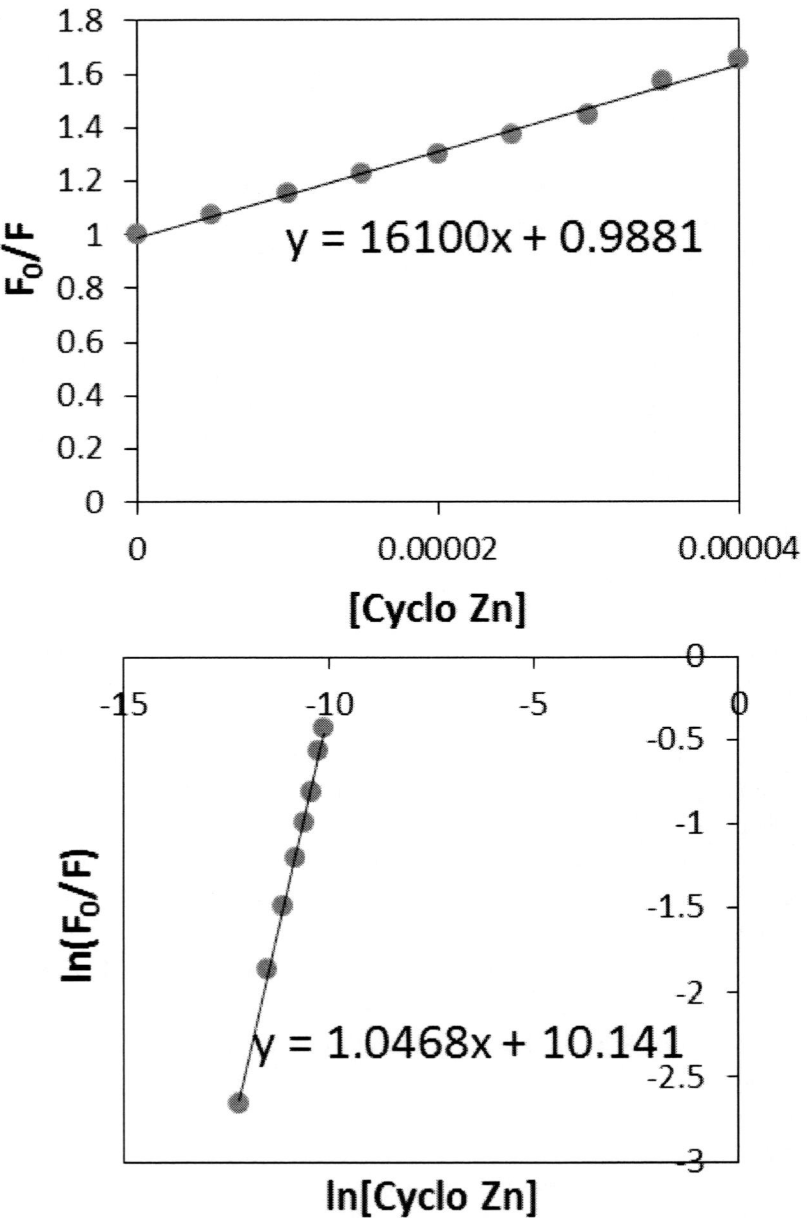

Figure 10. Stern-Volmer plot of **cyclo Zn**-Lysozyme composite system.

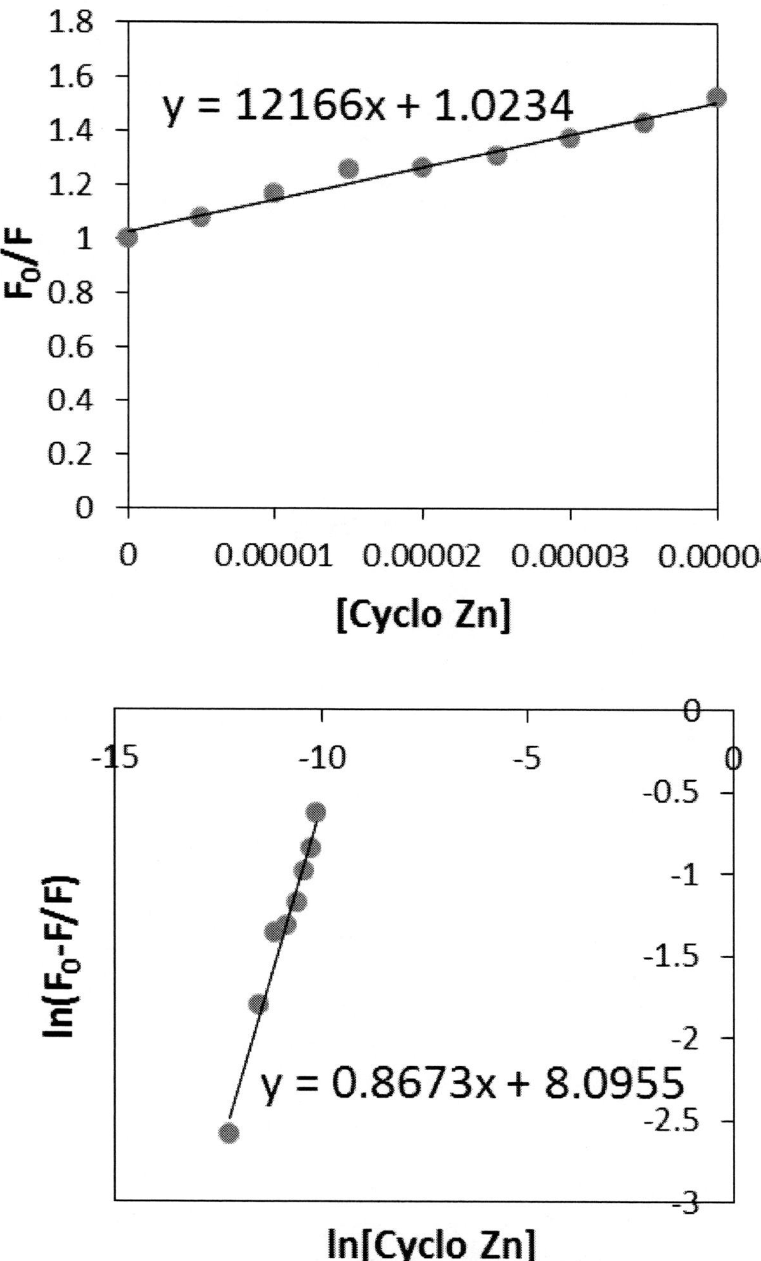

Figure 11. Stern-Volmer plot of **cyclo Zn**-HSA composite system.

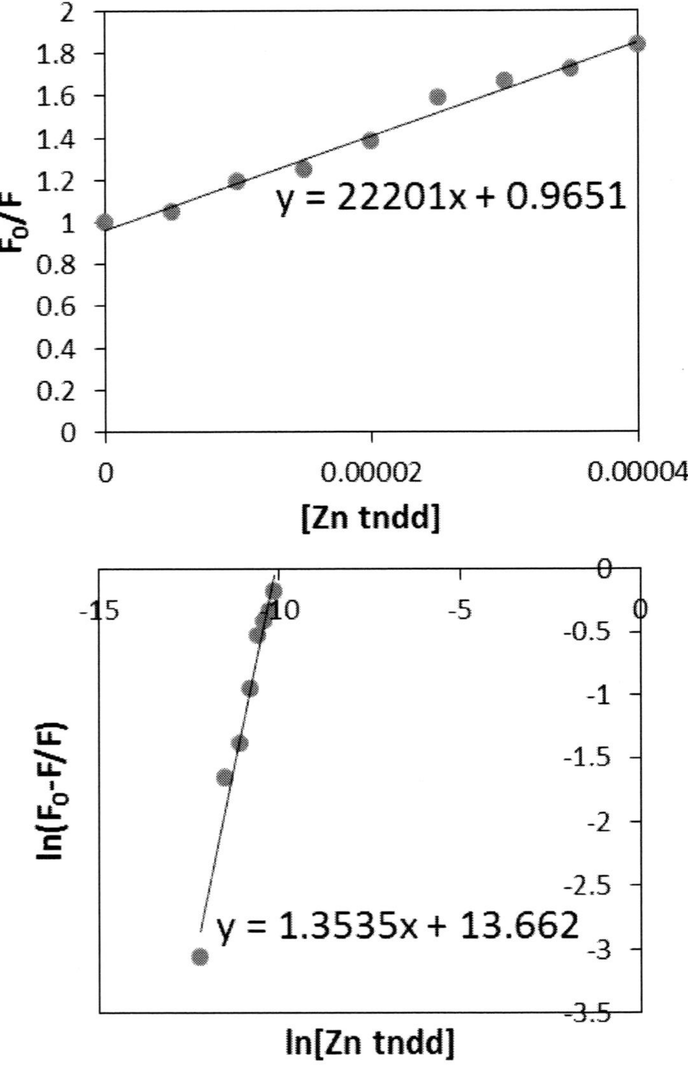

Figure 12. Stern-Volmer plot of **Zn tndd**-HSA composite system.

Similar to results of fluorescence spectra, thigh efficiency of energy transfer of this system could be confirmed numerically. As for binding constants, similar to Stern-Volmer constants, the **Zn tndd**-HSA composite system indicated large values, while **cyclo Zn**-lysozyme indicated larger values than **cyclo Zn**-HAS. Among them, the **Zn tndd**-HSA composite system exhibited the most strong interaction and the less energy loss during energy transfer.

Table 2. Parameters of each composite system (K_{SV}, K_{α}, n)

Composite systems	K_{SV} /M^{-1}	K_{α} /M^{-1}	n
cyclo Zn-lysozyme	1.61×10^4	2.533×10^5	1.0468
cyclo Zn-HAS	1.21×10^4	3.278×10^3	0.8673
Zn tndd-HSA	2.22×10^4	8.560×10^5	1.3535

CONCLUSION

Consequently, some metal complexes-protein composite systems could be obtained by using **cyclo Zn** and **Zn tndd**, which were confirmed by higher order structural changes detected with UV-vis and CD spectra. Quenching of fluorescence intensity on increasing concentration of metal complexes suggested energy transfer between metal complexes and proteins, in other words, docking due to intermolecular interactions. Stern-Volmer diagrams provided binding parameters, in which the numbers of binding-site (n) suggested the ratios of metal complexes: protein = 1:1 for each composite system. Among them, the **Zn tndd**-HSA composite system having the largest Stern-Volmer and binding constants was elucidated efficient energy transfer as well as strong intermolecular interactions. Additionally, metal complexes-protein composite systems appered intense emission even in polar solvents, in which intensity of emission are weaken commonly. The fact must be a merit to obtain advanced fluorescence materials.

REFERENCES

[1] Andruh, M., Costes, J.-P., Diaz, C., Gao, S., *Inorg. Chem.*,2009, 48, 3342-3359.

[2] Colacio, E., Ruiz, J., Mota, A. J., Palacios, M. A., Cremades, E.,Ruiz, E., White, F. J., Brechin, E. K., *Inorg. Chem.*, 2012,51, 5857-5868.

[3] Yan, P.-F., Lin, P.-H. Habib, F. Aharen, T. Murugesu, M. Deng,Z.-P. Li, G.-M., Sun W.-B., *Inorg. Chem.*, 2011, 50, 7059-7065.

[4] Colacio, E., Ruiz-Sanchez, J., White, F. J., Brechin, E. K., *Inorg. Chem.*, 2011, 50, 7268-7273.

[5] Jana, A., Majumder, S., Carrella, L., Nayak, M.,Weyhermueller, T.,

Dutta, S. Schollmeyer, D., Rentschler,E., Koner, R. Mohanta, S., *Inorg. Chem.*, 2010, 49, 9012-9025.

[6] Feltham, H. L. C., Clerac, R., Ungur, L. Vieru, V. Chibotaru,L. F., Powell, A. K., Brooker, S., *Inorg. Chem.*, 2012, 51,10603-10612.

[7] Margeat, O., Lacroix, P. G., Costes, J. P., Donnadieu, B.,Lepetit, C., *Inorg. Chem.*, 2004, 43, 4743-4750.

[8] Gomez, V., Vendier, L., Corbella, M., Costes, J.-P., *Inorg. Chem.*, 2012, 51, 6396-6404.

[9] Yang, X.-P., Jones, R. A., Wong, W.-K., Lynch, V., Oye, M.,M., Holmes, A. L., *Chem. Commun.*, 2006, 1836-1838.

[10] Burrow, C. E., Burchell, T. J. Lin, P.-H. Habib, P.,Wernsdorfer, W. Clerac, R., Murugesu, M., *Inorg. Chem.*, 2009, 48, 8051-8053.

[11] Wong, W.-K., Yang, X., Jones, R. A., Rivers, J. H., Lynch, V., Lo, W.-K., Xiao, D., Oye, M. M., Holmes, A. L., *Inorg. Chem.*, 2006, 45, 4340-4345.

[12] Pasatoiu, T. D., Madalan, A. M., Kumke, M. U., Tiseanu, C., Andruh, M., *Inorg. Chem.*, 2010, 49, 2310-2315.

[13] Whiteoak, C. J., Salassa, G., Kleij, A. W., *Chem. Soc.Rev.*, 2012, 41, 622-631.

[14] Bünzli, J.-C. G., Piguet, C., *Chem. Soc. Rev.*, 2005, 34, 1048-1077

[15] Okamoto, Y., Nidaira, K., Akitsu, T., *Int. J. Mol. Sci.*, 2011, 12, 6966-6979.

[16] Akitsu, T., Hirarsuka, T., Shibata, H., in *Magnets: Types, Usesand Safety*, Nova Science Publishers, 2012 and references therein.

[17] Hirarsuka, T., Shibata, H. Akitsu, T., in *Crystallography:Research, Technology and Applications*, Nova SciencePublishers, 2012 and references therein.

[18] Hutson, G. E., Dave, A. H., Rawal, V. H., *Org. Lett.*, 2007, 9, 3869-3872.

[19] Taylor, M. S, Zalatan, D. N., Lerchner, A. M., Jacobsen, E. N., *J. Am. Chem. Soc.*, 2005, 127, 1313-1317.

[20] Watanabe, E., Kaiho, A., Kusama, H., Iwasawa, N., *J. Am. Chem. Soc.*, 2013, 135, 11744-11747.

[21] Nocera, D. G., *Inorg. Chem.*, 2009, 48, 10001-10017.

[22] Ghiaci, M., Rezaei, B., Sadeghi., Safaei-Ghomi, J., *J. Chem. Eng. Data*, 2010, 55, 2792-2798.

[23] Valent, A., Melník, M., Hudecová, D., Dudová, B., Kivekäs, R., Sundberg, M. R., *Inorg.Chim. Acta*, 2002, 340, 15-20.

[24] Sallamn, S.A., Abbas, A.M., *J. Lumin.*, 2013, 136, 212-220.

[25] Fani, N., Bordbar, A. K., Ghayeb, Y., *Spectrochim. Acta Part A*, 2013, 103, 11-17.

[26] Ray, A., Seth, B. K., Pal, U., Basu, S., *Spectrochim. Acta Part A*, 2012, 92, 164-174.

[27] Cheng, L-X., Tang, J.-J., Luo, H., Jin, X.-L., Dai, F., Yang, J., Qian, Y.-P., Li, X.-Z., Zhou, B., *Bioorg. Med. Chem. Lett.*, 2010, 20, 2417-2420.

[28] Dehkordia, M. N., Bordbara, A.-K., Lincolnb, P., Mirkhani, Abdol-Khalegh., *Spectrochim. Acta Part A*, 2012, 90, 50-54.

[29] Dimiza, F., Papadopoulos, A., N., Tangoulis, V., Psycharis, V., Raptopoulou, C. P., Kessissogloua, D. P., Psomas, G., *Dalton Trans.*, 2010, 39, 4517-4528.

[30] van Vuong, Q., Siposova, K., Nguyen, T. T., Antosova, A., Balogova, L., Drajna, L., Imrich, J. Li, M. S., Gazova, Z., *Biomacromolecules*, 2013, 14, 1035−1043.

[31] Ancel, L., Gateau, C. Lebrun, C., Delangle, P., *Inorg. Chem.*, 2013, 52, 552−554.

[32] Marszalek, M., Konarska, A., Szajdzinska-Pietek, E., Wolszczak, M., *J. Phys. Chem. B* 2013, 117, 15987−15993.

[33] Peng, W., Ding, F., Peng, Y.-K., Jiang, Y.-T., Zhang, L., *Agric. Food Chem.*, 2013, 61, 12415-12428.

[34] Mohan, S., Kourentzi, K., Schick, K. A., Uehara, C., Lipschultz, C. A., Acchione, M., DeSantis, M. E., Smith-Gill, S. J., Willson, R. C., *Biochemistry*, 2009, 48, 1390–1398.

[35] Aureli, L., Cruciani, G., Cesta, M. C., Anacardio, R., Simone, L. D., Moriconi, A., *J. Med. Chem.*, 2005, 48, 2469-2479.

[36] Chen, J.-M., Ran, W.-J., Gao, F., Duan, R., Zhang, Y.-H., Zhu, Z.-A., *J. Coord. Chem.*, 2007, 60, 2485-2497.

[37] Mihara, H., Xu, Y., Shepherd, N. E., Matsunaga, S., Shibasaki, M., *J. Am. Chem. Soc.*, 2009, 131, 8384-8385.

[38] Knudsen, K. R., Risgaard, T., Nishiwaki, N., Gothelf, K. V., Jørgensen, K. A., *J. Am. Chem. Soc.*, 2001, 123, 5843-5844.

[39] Pasatoiu, T. D., Tiseanu, C., Madalan, A. M., Jurca, B., Duhayon, C., Sutter, J. P., Andruh, M., *Inorg. Chem.*, 2011, 50, 5879–5889.

[40] Akine, S., Utsuno, F., Taniguchi, T., Nabeshima, T., *Eur. J. Inorg. Chem.*, 2010, 20, 3143-3152.

[41] McCarthy, P. J., Hovey, R. J., Ueno, K., Martell, A. E., *J. Am. Chem. Soc.* 1955, 77, 5820-5824.

[42] Das, L. K., Biswas, A., Gómez-García, C. J., Drew, M. G. B., Ghosh, A., *Inorg. Chem.* 2014, 53, 434-445.

[43] Xu, Y., Lin, L., Kanai, M. Matsunaga, S., Masakatsu, S., *J. Am. Chem. Soc.*, 2011, 133, 5791-5793.

[44] Yoshino, T., Morimoto, H., Lu, G., Matsunaga, S., Shibasaki, M., *J. Am. Chem. Soc.*, 2009, 131, 17082-17083.

[45] Suga, H., Kakehi, A., Mitsuda, M., *Bull. Chem. Soc. Jpn.*, 2004, 77, 561-568.

[46] He, L.-L., Wang, X., Liu, B., Wang, J., Sun, Y.-G., *J. Sol. Chem.*, 2012, 41, 1853–1865.

[47] Fard, F. J., Khoshkhoo, Z. M., Mirtabatabaei, H., Housaindokht, M. R., Jalal, R., Hosseini, H. E., Bozorgmehr, M. R., Esmaeili, A. A., Khoshkholgh, M. J., *Spectrochim. Acta Part A*, 2012, 97, 74–82.

In: Threonine
Editor: Jacob Coleman

ISBN: 978-1-63482-554-2
© 2015 Nova Science Publishers, Inc.

Chapter 4

SPECTROSCOPIC AND ELECTROCHEMICAL STUDIES ON METALLOPROTEIN (LACCASE) AND CU(II) COMPLEX MEDIATORS AS MODEL SYSTEMS FOR BIOFUEL CELL CATHODES

Yu Kurosawa, Erika Tsuda, Masahiro Takase, Nanami Yoshida, Yuto Takeuchi, Yuya Mitsumoto and Takashiro Akitsu*

Department of Chemistry, Faculty of Science,
Tokyo University of Science, Tokyo, Japan

ABSTRACT

Serine, threonine, and alanine are majorly contained amino acid residues in laccase, a metalloprotein which reduces oxygen into water. In which threonine may potentially play a role in intermolecular interaction toward ligands of small molecules and catalytic group. Laccase are employed in typical biofuel cell cathodes with a mediator metal complex such as ferrocene.

---

* Corresponding author: Takashiro Akitsu: Department of Chemistry, Faculty of Science, Tokyo University of Science, 1-3 Kagurazaka, Shinjuku-ku, Tokyo 162-8601, Japan. Tel.: +81-3-5228-8271, fax: +81-3-5261-4631, E-mail address: akitsu@rs.kagu.tus.ac.jp.

In this paper, we report on oxygen reduction by laccase with other metal complexes known electron mediators in acetate buffer suspension and in carbon paste electrodes. Furthermore, in order to develop low-cost mediators, we prepared and tested some Cu(II) complexes, namely [Cu (phen)$_2$]$^{2+}$, [Cu(phen-derivative)$_2$]$^{2+}$, and [Cu(Schiff base)$_2$], which were characterized by means of elemental analysis, IR, UV-vis, and CD spectroscopy. Hybrid systems of the complexes and laccase were also prepared for comparison current and potentials of oxygen reduction.

INTRODUCTION OF BIOFUEL CELLS

Generally, the lifetimes of fossil fuels are limited, for example, oil, natural gas, and coal is 40, 60, and 200 years, respectively. Various technologies have been developed as alternative energy especially we focus on biofuel cells in this chapter. Biofuel cells generate electricity (flow of electron to external circuit as energy from negative electrodes) by organic matter oxidized by enzyme or microbial cells. In principle, it is clean devices without increasing CO_2, because they merely exchange organic matters to electric power using certain catalysts. Among several types of fuel cells, biofuel cells are consisted of enzymes (protein of biocatalysts) and work under mild conditions, namely room temperature and ambient pressure.

Theoretically, biofuel cells react to generate electric power by the following reactions.

Anode: $C_6H_{12}O_6 + 6O_6 + 6H_2O \rightarrow 6CO_2 + 12H_2O$

Cathode: $O_2 + 4H^+ + 4e^- \rightarrow 2H_2O$

Evaluation of oxidation-reduction (redox) potentials of substrates and products is important for performance as battery. As for glucose ($C_6H_{12}O_6$) as substrate, which are oxidized to CO_2 at the anode. On the other hand, O_2 is simultaneously reduced to water at the cathode. These redox potentials are -0.42, +0.82 V respectively, therefore the electric potential difference is +1.24 V [1]. Since 1 mol of glucose generate 24 mol of electron, Faraday's law of electrolysis provides the resulting quantity of electricity from 1 g of glucose for an hour is 3.57 Ah (24 × 9.6485 × 104 / (3600×180) = 3.57), of which electric energy is 1.8 Wh. Compared with the similar electric energy of nickel metal hydride battery (3.9 Wh) indicating high energy density, biofuel cells are suitable for power source of mobile electric appliances.

In addition, the substrate of biofuel cells in the solid state devices may be safer than those of conventional fuel cells using metal catalysts (for example hydrogen or methanol).

In this way, various types of fuel cells (Figure 1) have both merits and weak points. The first biofuel cells containing enzyme was reported in 1964 [2] and fundamental studies on them have mainly been continued so far to improve low electric power generation. Some enzyme biofuel cells generating electric power from abundant biofuel are composed of oxidoreductase [3, 4], whose advantage is simple structure (possibility for easy miniaturization), safety, metal or inorganic catalyst free, and low environmental burden.

On the other hand, there are many serious problems about biofuel cells at present. At present, theoretical conversion efficiency of electron from glucose may be less than 10 %, in which a number of enzymes found in metabolism systems are used without matching each other. Furthermore, stability, especially, durability of enzyme may also be problems, which should be overcome for practical application. For example, durability of enzyme against pH is improved by gene recombination [5, 6]. New immobilization method or immobilization of enzyme in sol-gel matrix maintaining biocompatibility may be expected to improve durability for electric transfer directly from enzyme electrode. As for activity as catalytic systems, gold nano-particles or some carbon nano-materials are expected to improve electron transfer when they bind to enzyme using high surface with porous structures of enzymes. As for cell voltage and output power, new design of biofuel cells (of course maintaining biocompatibility) has been developed from the view points of materials chemistry or chemical engineering.

Among such problems and improvement, "mediators" are employed to enhance electron transfer between enzymes and electrodes (bioanodes or biocathodes) regardless of surface area restriction.

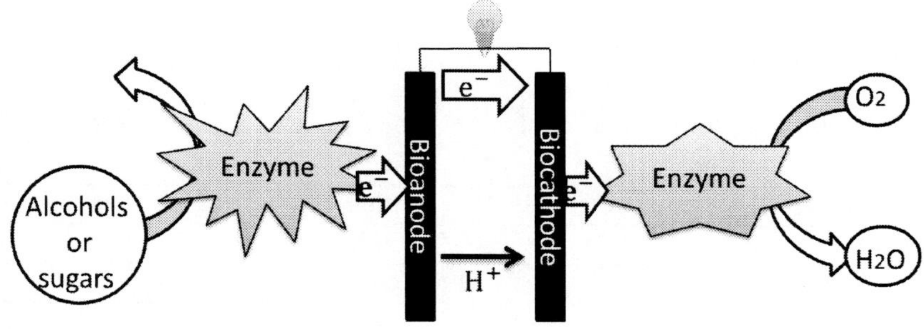

Figure 1. Schematic reaction mechanism of typical biofuel cells.

Kinetics of electron transfer about energy conversion dominates current, because mediators may be key component to improve the efficiency of biofuel cells. In biochemistry, mediator can exchange electrons if metabolic degradation of organic matters proceeds sufficiently. Appropriate mediators indicate similar redox potentials to enzymes (Figure 2, Table 1) [7]. This is molecular design strategy for mediators to avoid low rates of electron transfer as well as proton transfer from anode to cathode or diffusion rates of substrates and oxidant in electrodes, and consequently resulting current density.

In the last section of this paper, our conceptual development of mediators using transition metal complexes without expensive metals will be reported. To adjust proper redox potentials, not only metals but also organic ligands should be designed.

Transition metal complexes are impossible to be included inner space of enzymes, while have advantage to conduct current efficiently. Whereas metal nano-particles are possible to be included inner space of enzyme selectively (better for spatial reasons), while it is hard for them to carry much current.

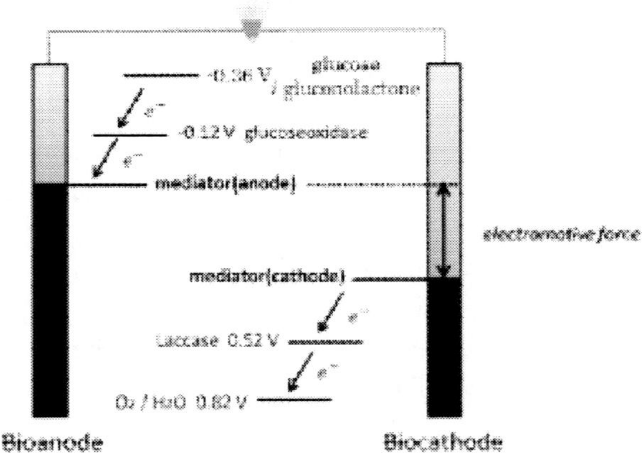

Figure 2. Appropriate redox potentials required for mediators.

Table 1. Redox potentials (vs SCE) for various mediators [8, 9]

mediators	Redox potentials (V)	mediators	Redox potentials (V)
ferrocene	0.2	ABTS	0.65
Fe-porphyrin	0	$[Fe(CN)_6]^{n-}$	0.5
$[Os(bpy)_2Cl_2]^{n+}$	0.1	p-quinone	0.25
$[Ru(bpy)_2Cl_2]^{n+}$	0.3		

Therefore, we focused on metal complexes as target candidates for suitable mediators.

Moreover, other methodologies have been applied to prepare electrodes. Fixing mediators and enzyme on electrode surface is typical method to enhance electron transfer associated with electrochemical reactions. From this viewpoint, it was reported that fixed polymers binding to osmium complexes and multicopper oxidase (MCO), reducing O_2 to water, exhibit a few mA cm^{-2} current density [10]. In this system, not only dissolved oxygen but also oxygen gas can be catalyzed by this cathode of a biofuel cell.

LACCASE AS A CATALYTIC ENZYME

Laccase, one of the members of MCO family, has attracted a great deal of attention because of that utility of biofuel cells. The reason is because laccase including four copper active sites can catalyze 4-electron reduction of oxygen except to yield water-soluble peroxide as intermediates. All three types of copper active sites are including into laccase. Depending on spectral properties, they can be classified into so-called type 1, type 2, and type 3.

Type 1 copper (also called blue copper) showing blue color due to strong ligand to metal charge transfer absorption at about 600 nm by coordination of cysteine residues, and they are generally contained in electron transfer metalloproteins. Their electronic spectra exhibit characteristic band from UV region to NIR region besides d-d bands.

ESR spectra of type 1 copper exhibit hyperfine coupling, owing to afford a twisting tetrahedral coordination geometry (by His, His (imidazole-N), Met (S), and Cys (S) typically to be T_d symmetry or also by additional peptide main chain to be extended C_{3v} symmetry ideally) as well as mixed valence between Cu(II) and Cu(I) states. Due to the features, redox potential is relatively high value (0.2-0.8 V vs NHE).

Type 2 copper (namely normal copper sites) does not appear such characteristic color like type 1 and takes a mononuclear coordination environment (by imidazole-N and COO$^-$-O) with various geometries.

Type 3 copper is binuclear active centers of phenol-O bridged diamagnetic Cu(I)-Cu(I) or antiferromagnetically coupled Cu(II)-Cu(II) sites (ESR silent), whose magnetic or spectroscopic properties are affected by bond lengths and angles of the binuclear moieties [10, 11], in other words, electronic states depend on steric factors.

As shown in Figure 3, from surface of protein molecules, electrons are provided from (potentially oxidized) substrates are received by type 1 copper and transferred to the trinuclear cluster of type 2 and type 3 about 1.3 nm distance from the type 1 site, which catalyzes 4-electron reduction of oxygen (important process of redox reaction) without forming O^{2-}, H_2O_2, OH^- finally. In this way, laccase is called multicopper oxidase (MCO).

One of the most important factors of these redox reactions may be reaction rate. MCO can react more effectively than copper efflux oxidase (indicating photocatalytic electrode activity) or inorganic Pt catalyst (indicating overpotential) at room temperature, ambient pressure, and neutral pH conditions. Though manifold MCO perform in weak acidity condition, proper pH for manifold enzyme is almost neutral. Neutral condition is necessary for final target of biofuel cells *in vivo*. Not all metalloproteins can be investigated directly by means of electrochemical methods, because active site is far from surface of electrodes and slow reactions of electron transfer.

METAL COMPLEXES PLAYING BIOINORGANIC ROLES

Traditionally, determination of molecular structures is the most important aspects of studies on coordination chemistry.

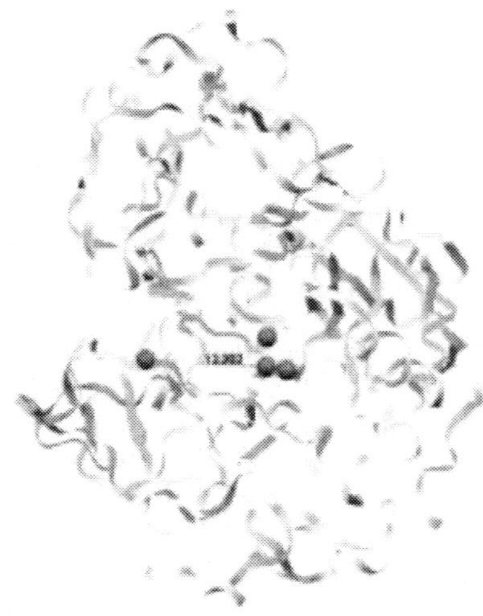

Figure 3. Structure of laccase with copper active sites.

Structures of chlorophyll are well investigated because of their importance in photosynthesis associated with plants. Besides molecular structures, these molecules form supramolecular assemblies of characteristic arrangements to concentrate light. Their structural studies are in progress at present [12].

Natural chlorophylls contain manganese (Figure 4), while zinc substituted ones are also prepared as attractive model compounds due to their stability [13]. A lot of design of such complexes has reported and applied to various field such as computational chemistry [14], application for dye sensitized solar cells [15].

As for synthetic metal complexes, for example, some binuclear copper complexes (Figure 5) indicate interesting catalytic properties (expected as cooperative reactions) as well as magnetic properties of given spin states [16], which cannot be observed for mononuclear copper complexes [17, 18]. Since they form complicated systems, suitable model compounds should be prepared for detailed studies.

Copper is one of the most common metal elements like iron or zinc. Various valence states, Cu(I), Cu(II), and Cu(III) can be found in (electron transferring) metalloproteins containing copper. As Lewis acids, Cu(I), Cu(II), and Cu(III) are soft, middle, and hard acids, respectively.

According to HSAB theory, Cu(I) tends to coordinate to soft bases such as sulfur ligands. Coordination numbers range from 2 to 12, while 4, 5, 6-coordinated complexes are found widely.

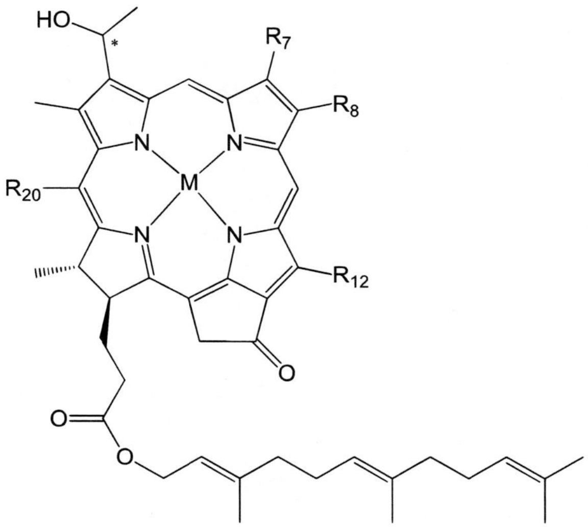

Figure 4. Typical structure of chlorophyll.

A Cu(I) complex commonly affords a tetrahedral coordination geometry, while a Cu(II) complex flexibly affords square planar, trigonal bypiramidal, or distorted octahedral coordination geometries. In actual copper proteins exhibiting redox reactions, structural features can be controlled by valence states as well as steric restriction by peptide chains.

The affinity of complexes to residue of amino acid is classified as follow roughly.

Asp, Glu (-COO$^-$) : Mg^{2+}, Ca^{2+}, Mn^{3+}, Fe^{3+}, Fe^{2+}, Zn^{2+}
His (imidazole N) : Fe^{2+}, $\underline{Cu^{2+}}$, $\underline{Cu^{+}}$, Mn^{2+}, Zn^{2+}
Cys (-S$^-$) : Zn^{2+}, $\underline{Cu^{+}}$, $\underline{Cu^{2+}}$, Fe^{3+}, Fe^{2+}, $Mo^{4\sim6+}$, $Ni^{1\sim3+}$
Met (-SCH$_3$) : Fe^{2+}, Fe^{3+}, $\underline{Cu^{+}}$, $\underline{Cu^{2+}}$
Tyr (-C$_6$H$_4$O$^-$) : Fe^{3+}, $\underline{Cu^{2+}}$

Bioinorganic chemistry deals with Werner type complexes, while bioorganometallics deals with non-Werner type ones having not lone-pair donating coordination M-L bonds between Lewis acid (M) of metal and Lewis base (L) of ligands but covalent M-C(carbon) bonds. Various structures of them have been also determined by means of X-ray crystallography.

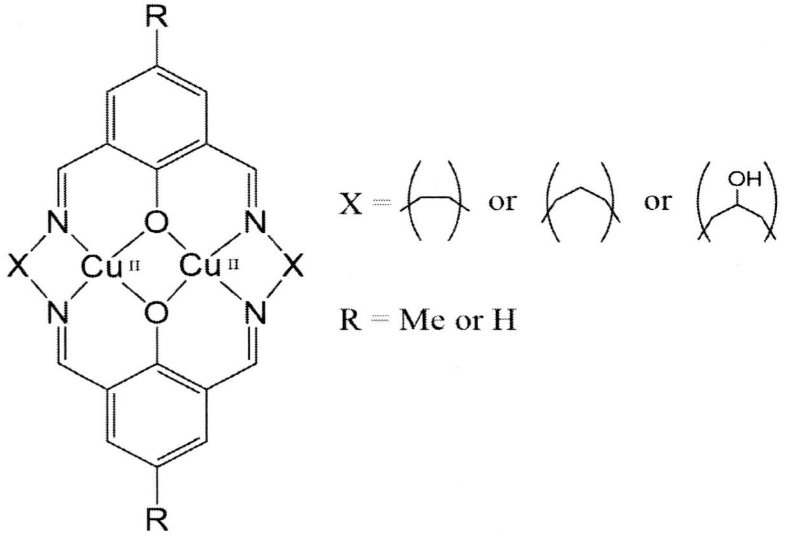

Figure 5. Functional binuclear copper complexes.

Structures of complexes lead to expect their properties, hence structural determination is important. Indeed, ferrocene was studied by Wilkinson and Fischer since early 1952 [19].

DOCKING OF METAL COMPLEXES AND PROTEINS

Metal complexes, which have various structures, show multifarious biochemical behavior [20]. Metal complexes can also be useful as a homogeneous catalysis similar to biocatalysts not only sole species but also hybrid catalysts with biopolymers (proteins or DNA), which is called artificial metalloenzymes or 'Artzymes'. Artificial metalloenzymes indicate high selectivity for substrate by designing ligands or combination with components, though they are weak against heat. Their functions can be applied organic reactions such as hydrogenation, bennzannulation and Diels-Alder [21-23].

To date, docking of metal complexes into protein has been studied, though accurate binding features have not been elucidated clearly.

However, development of protein-ligand docking simulation software, for example GOLD or Protein DF, enables docking simulation, and becomes clear about accurate bind formation to discuss catalytic activity [24] based on information proteins-ligands or proteins-proteins [25].

PLELIMINARY STUDY ON NEW COPPER MEDIATORS

In this section, we report on testing some complexes as mediators and preliminary results of electrochemical and spectroscopic properties. Commonly, ferrocene or osmium complexes (Figure 6) binding to polymers are used for this purpose. The source of osmium metal is expensive relatively.

Organic/inorganic hybrid materials composed of metal complexes and polymers have potential to emerge various functions, for example, exhibiting photochromism by containing azo-groups [26, 27], emission [28], or chirality [29]. Herein we expand their application to the cathode of biofuel cells of mediators and laccase to improve electron transfer between electrode and enzyme. Although osmium or ruthenium complexes are usually employed as mediator, they are commonly expensive. Then, we prepared some copper(II) complexes, **1**, **2**, **3**, and **4** according to the literatures [30-33], and ferrocene **5** (Figure 7) expected as low costs and abundant resources and tested them.

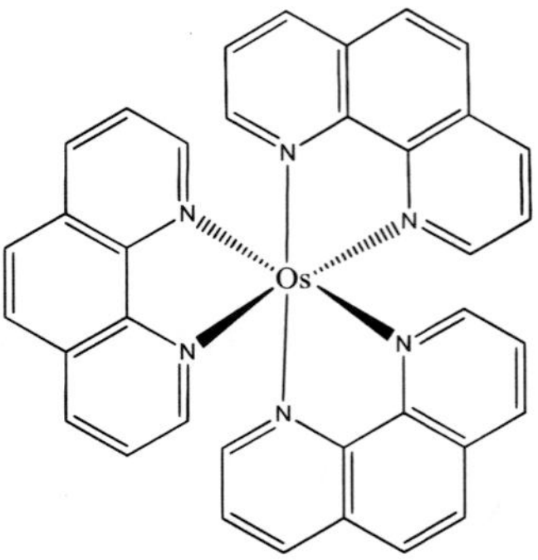

Figure 6. Example of an osmium complex as a mediator. Valence states are omitted for clarity.

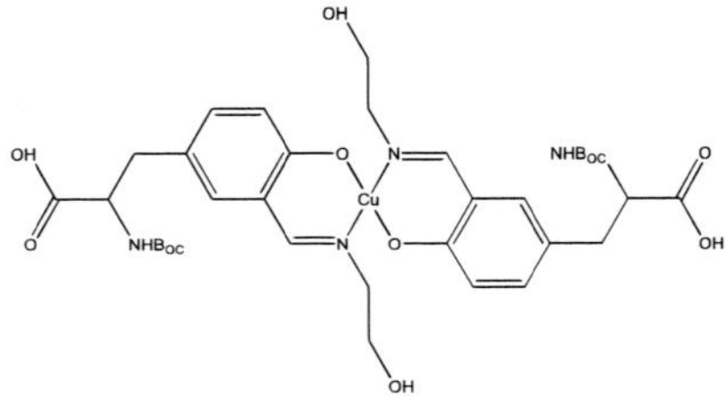

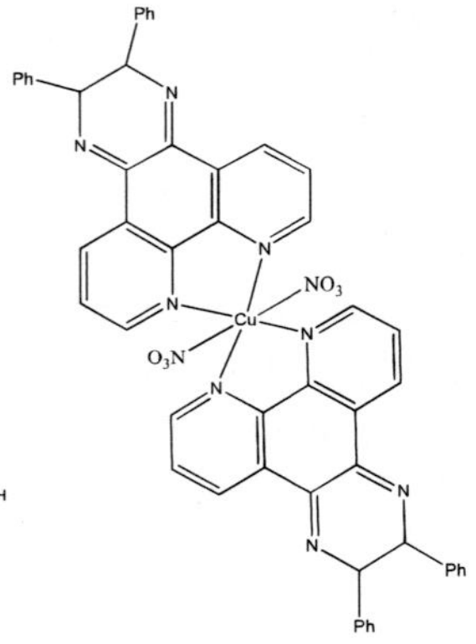

Figure 7. Complexes **1-5** tested as mediators.

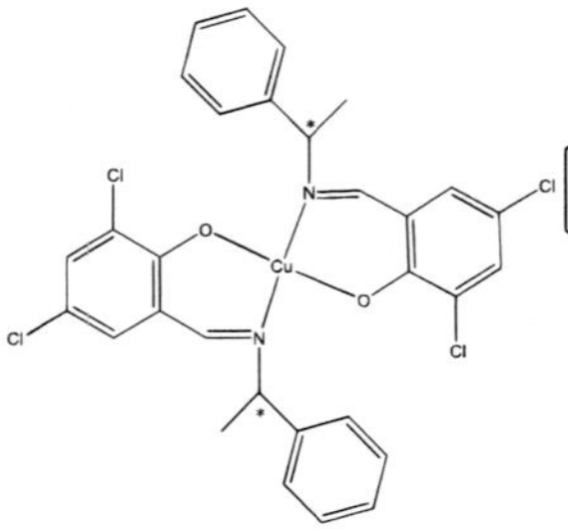

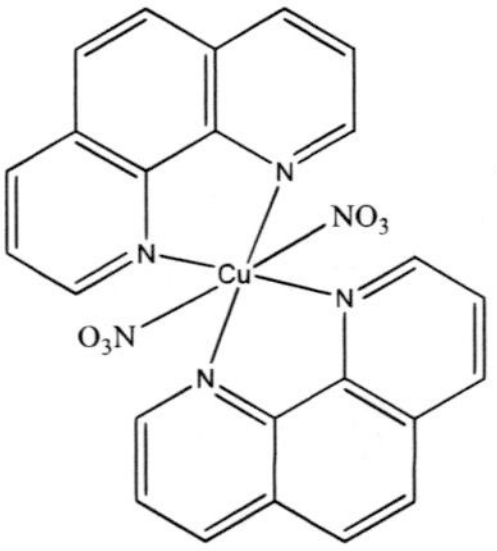

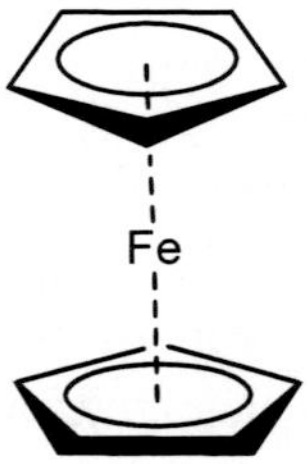

Complexes **1-5** have been investigated as the mediators to laccase as hybrid systems [34]. Complex **4** is bis(1,10-phenentholorine)copper(II) cation, which is typical example of copper(II) complexes. Besides copper(II) complexes, only complex **5** is ferrocene (iron (II/III) complex) which is used a reference substance for redox reactions commonly. In acetate buffer (pH 4.5, 50 mM), UV-vis spectra appeared predominant peaks at 220, 340, and 360 nm for **1**, 212, 252, 285, 365 nm for **2**, 211, 240, 280, 304, and 380 nm for **3**. As for chiral complexes, CD spectra appeared predominant peaks at 220, 340, and 360 nm for **1**, 233, 256, 300, 360, and 400 nm for **3**.

CV (cyclic voltammetry) (in DMSO with 0.1 M TBAP, at 100 mV/s) exhibited oxidation/reduction potentials of Cu(II)/Cu(I) at 0.42/0.18 V for **1**, -0.16/-0.26 for **2**, -0.12, -0.36, -0.54/ -0.16, -0.44, -0.68 V for **3**, -0.14/0.24 for **4** (and for comparison, 0.25/0.15 V for **5**) (Figure 8). All redox reactions are reversible, and only **3** exhibited multi-step reaction accompanying with organic ligands including halogen.

On the other hand, CV data were also measured using carbon paste electrode in order to apply for metal complexes-laccase hybrid systems, in which laccase was supplied commercially. In acetate buffer (pH 4, 200 mM, at 50 mV/sec), CV (Figure 9) exhibited oxidation/reduction potentials at 0.3/-0.4 V (5 μA) for laccase (as control), 0.1, 0.3, 0.5/-0.3, -0.1 V for **1**, 0.3/-0.1 for **2**, 0.1/-0.3 for **3**, 0.3, 0.0/0.1, -0.1 for **4**, and 0.4/0.2 for **5**. Only **1** and **4** exhibited multi-step reactions in the conditions. As for metal complexes-laccase hybrid systems whose potential indicated negative shifts, observed current values (potentials) were 27 μA (-0.3 V) for **1**+laccase, 10 μA (-0.3 V) for **2**+laccase, 100 μA (-0.3 V) for **5**+laccase. No current was observed for **4**+laccase.

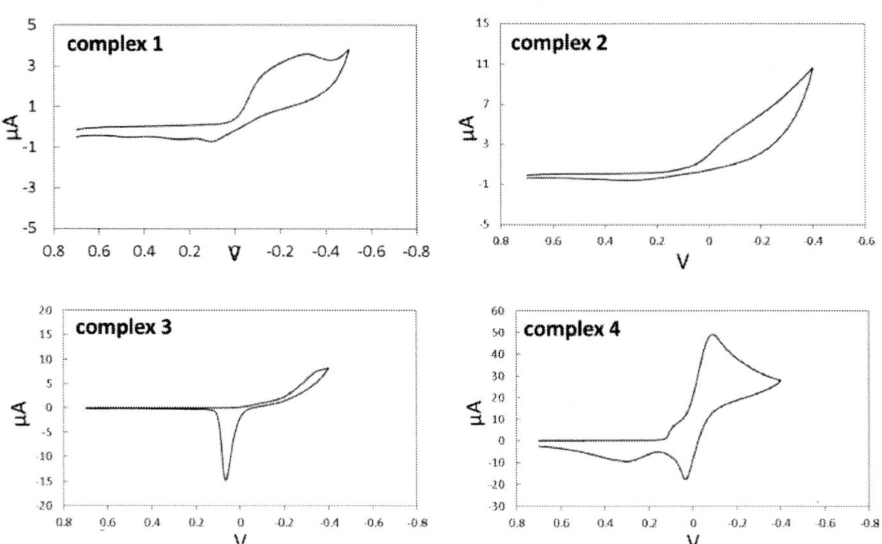

Figure 8. CV data for complexes **1-4**.

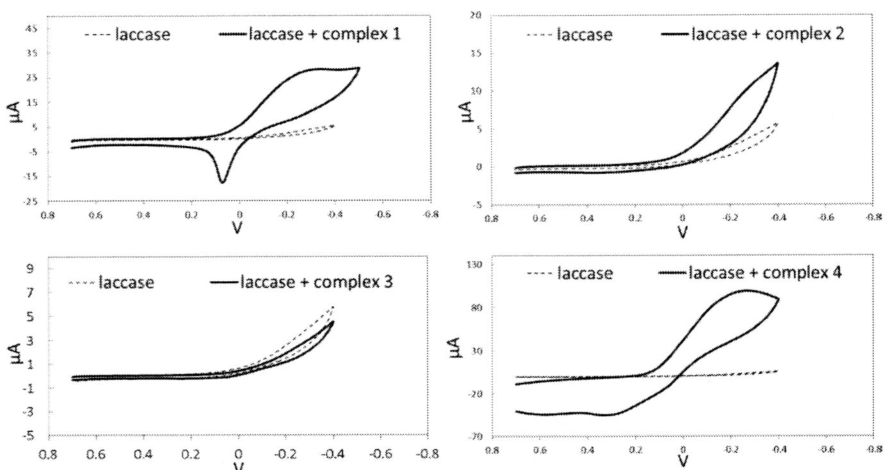

Figure 9. CV data for hybrid systems of **1-4**+laccase.

Consequently, compared with control value of 115 µA (0.1 V) for **5**+laccase, **4**+laccase may be the best mediator systems among the present copper(II) complexes (Figure 9). Correlation between geometrical or electronic structures (binding to proper sites or residues of proteins by molecular recognition) and electrochemical functions (especially large current for a good mediator) will be elucidated in future.

REFERENCES

[1]   H. Sakai, T. Nakagawa, Y. Tokita, T. Hatazawa, T. Ikeda, S. Tsujimura, K. Kano, *Energy Environ. Sci.*, 2009, 2, 133.

[2]   M. J. Cooney, V. Svoboda, C. Lau, G. Martin, S. D. Minteer, *Energy Environ. Sci.*, 2008, 1, 320, and references therein.

[3]   K. P. Prasad, Y. Chen, P. Chen, *ACS Appl. Mater. Interfaces*, 2014, 6, 3387.

[4]   M. J. Moehlenbrock, S. D. Minteer, *Chem. Soc. Rev.*, 2008, 37, 1188.

[5]   N. Yuhashi, M. Tomiyama, *Biosensors and Bioelectronics*, 2005, 15, 2145.

[6]   N. L. Akers, C. M. Moore, S. D. Minteer, *Electrochimica Acta*, 2005, 50, 2521.

[7]   J. Kulys, T. Buchrasmussen, K. Bechgaard, V. Razumas, J. Kazlauskaite, J. Marcinkeviciene, J. B. Christensen, H. E. Hansen, *J. Mol. Catal.*, 1994, 91, 407.

[8] F. P. Cardoso, S. A. Neto, L. B. Crepuldi, S. Nikolaou, V. P. Barros, A. R. De Andrade, *J. Electrochem. Soc.*, 2014, 161, F445.

[9] W. Nogala, K. Szot, M. Burchardt, F. Roelfs, J. Rogalski, M. Opallo, G. Wittstock, *Analyst*, 2010, 135, 2051.

[10] A. Heller, *Curr. Op. Chem. Biol.*, 2006, 10, 664.

[11] T. R. Ralph, M. P. Hogarth, *Platinum Metals Rev.*, 2002, 46, 3.

[12] N. Wakao, N. Yokoi, N. Isoyama, A. Hiraishi, K. Shimada, M. Kobayashi, H. Kise, M. Iwaki, S. Itoh, S. Takaichi, *Plant Cell. Physiol.*, 1996, 37, 889.

[13] G. S. Orf, R. E. Blankenship, *Photosynth. Res.*, 2013, 116, 315.

[14] T. Yamamura, S. Suzuki, T. Taguchi, A. Onodera, T. Kamachi, I. Okura, *J. Am. Chem. Soc.*, 2009, 131, 11719.

[15] Y. Yang, R. Jankowiak, C. Lin, K. Pawlak, M. Reus, A. R. Holzwarth, J. Li, *Phys. Chem. Chem. Phys.*, 2014, 16, 20856.

[16] K. Das, S. Nandi, S. Mondal, T. Askun, Z. Canturk, P. Celikboyun, C. Massera, E. Garribba, A. Datta, C. Sinha, T. Akitsu, *New J. Chem.*, 2015, 39, 1101.

[17] C. Yang, M. Vetrichelvan, X. Yang, B. Moubaraki, K. S. Murray, J. J. Vittal, *Dalton Trans.*, 2004, 1, 113.

[18] L. K. Thompson, S. K. Mandal, S. S. Tandon, J. N. Bridson, M. K. Park, *Inorg. Chem.*, 1996, 35, 3117.

[19] G. Wilkinson, M. Rosenblum, M. C. Whiting, R. B. Woodward, *J. Am. Chem. Soc.*, 1952, 74, 2125.

[20] A. Ray, B. K. Seth, U. Pal, S. Basu, *Spectrochim. A*, 2012, 92, 164.

[21] E. Sansiaume-Dagousset, A. Urvoas, K. Chelly, W. Ghattas, J. Marechal, J. Mahy, R. Ricoux, *Dalton Trans.*, 2014, 43, 8344.

[22] J. R. Carey, S. K. Ma, T. D. Pfister, D. K. Garner, H. K. Kin, J. A. Abramite, Z. Wang, Z. Guo, Y. Lu, *J. Am. Chem. Soc.*, 2004, 126, 10812.

[23] J. Steinreiber, T. R. Ward, *Coord. Chem. Rev.*, 2008, 252, 751.

[24] T. Ueno, T. Koshiyama, M. Ohashi, K. Kondo, M. Kono, A. Suzuki, T. Yamane, Y. Watanabe, *J. Am. Chem. Soc.*, 2005, 127, 6556.

[25] T. Ueno, T. Koshiyama, S. Abe, N. Yokoi, M. Ohashi, H. Nakajima, Y. Watanabe, *J. Organometal. Chem.*, 2007, 692, 142.

[26] K. Jensen, G. Panagiotou, I. Kouskoumvekaki, *PLOS Comput. Biol.*, 2014, 10, 1.

[27] O. A. Blackburn, B. J. Coe, J. Fielden, M. Helliwell, J. J. W. McDounal, M. G. Hutchings, *Inorg. Chem.*, 2010, 49, 9136.

[28] A. Khandar, Z. Revani, *Polyhedron*, 1998, 18, 129.

[29] C. Kominato, T. Akitsu, *Lett. Appl. NanoBioSci.*, in press.

[30] M. E. Jung, T. I. Lazarova, *J. Org. Chem.*, 1997, 62, 1553.

[31] W. Paw, R. Eisenberg, *Inorg. Chem.*, 1997, 36, 2287.

[32] Y. Einaga, T. Akitsu, *Polyhedron*, 2005, 24, 2933.

[33] K. J. Catalan, S. Jackson, J. D. Zubkowski, *Polyhedron*, 1995, 14, 2165.

[34] Y. Kurosawa, T. Akitsu, unpublished data (presented in *The 95rd Annual Meeting of The Chemical Society of Japan*, 2015).

INDEX

bloodstream, 44
body composition, 45
Brazil, 27, 29
Brevibacterium flavum, vii, 2, 4, 5, 7, 9, 15, 16, 17, 18, 22, 23, 25

C

Ca^{2+}, 80
calcium, 32
calibration, 65
candidates, 77
carbon, ix, 7, 10, 22, 24, 25, 74, 75, 80, 83
catalysis, 50, 81
catalyst, 75, 78
catalytic activity, 81
catalytic properties, 79
catalytic system, 75
cell culture, vii, 2
cell death, 43
central nervous system, 28
chemical, 6, 7, 8, 30, 50, 75
chemical reactions, 50
classification, 18
CO_2, 74
collagen, 2, 28, 38, 39, 43
commercial, 4, 6, 28, 46, 51
complement, 11
composites, 61
composition, 7, 10, 30, 32
compounds, 6, 50, 79
computer simulations, 57
conversion rate, 22
cooling, 51, 52
coordination, 50, 77, 78, 80
copper, 77, 78, 79, 80, 81, 82, 84
corepressor, viii, 2
crystal structure, 51
cultivation, 9, 19, 20, 22, 23
cultivation conditions, 23
culture, viii, 2, 7, 8, 9, 10, 11, 15, 16, 22, 25
culture conditions, 25
culture medium, viii, 2
cysteine, 77
cystine, viii, 28, 31, 32

D

deformation, ix, 50
degradation, 19, 76
degradation process, 19
denaturation, 11
dendrogram, viii, 2
deposition, 38, 39, 40, 43, 47
diet, 30, 31, 35, 38, 40, 41, 43, 44, 45, 46
diffuse reflectance, 52, 53, 54, 55, 56
diffusion, 76
diffusion rates, 76
digestion, 2, 28
digestive enzymes, 3
dissolved oxygen, 77
distilled water, 10
DNA(s), 5, 8, 10, 11, 25, 26, 51, 56, 81
DNA repair, 26
drugs, 9, 50

E

E.coli, 5
efficiency of use, 40
egg, vii, viii, ix, 28, 29, 31, 32, 33, 34, 35, 36, 37, 41, 42, 43, 45, 46, 49, 51
egg white, vii, ix, 45, 49
electrodes, ix, 74, 75, 76, 77, 78
electrolysis, 74
electron(s), ix, 50, 74, 75, 76, 77, 78, 79, 81
electronic structure, 84
electrophoresis, 10, 11
emission, ix, 49, 50, 51, 56, 57, 69, 81
energy, ix, 30, 32, 43, 50, 63, 65, 68, 69, 74, 76
energy density, 74
energy transfer, ix, 50, 63, 65, 68, 69
engineering, 5, 24, 25, 75
environment, 29, 50, 77
environmental factors, 28
enzyme(s), 4, 6, 8, 24, 74, 75, 76, 77, 78, 81
ESR spectra, 77
essential amino acids, vii, ix, 1, 9, 19, 49
excitation, 57, 65